TIP TO BOOST CLIMATE RESILIENCE

HONORE MATAMBA M. LUKASU

The Eagle Literary, LLC
1401 21st Street STE R Sacramento, California 95811
theeagleliterary.com
(209) 692-5191

Font Used: EB Garamond is a serif typeface designed by Georg Duffner and available through Google Fonts. It is licensed under the SIL Open Font License.

ISBN (Hardback): 979-8295692185
ISBN (Paperback): 979-8295611131
ISBN (Ebook): 979-8295611148

Library of Congress Control Number: 2026903872

Printed in the United States of America

Table of Content

Acknowledgements

A finger alone can not pick up better a louse, they say.

For this book to take the shape you now hold in your hands, this saying was a reality.

Thanks to the Lord who keeps me alive, protects me and opened my mind to get to do what I could to get this book become a reality.

In addition to the protection and grace of God, the moral supports from my family and friends was a necessity for me.

May my extended biological family find deep gratitude. I quote here Celestin Kabange Tshibamba, Astrid Kayowa Kabanga, Josephine Masengu, Jeanpy Tshimanga, Jean Mbuyi Sabwa and Marte Ngalula.

May the souls of my dad late David Katomba wa Ntumba, mom late Bernadette Kaseka Nsanza, my elder brothers late Crispin (Tom) Mutombo Kazadi Kolesha, late Jeanpierre Ntumba Kazadi, late David Mbuyi Katomba, my little brother late Kabanga Nzengu and my twin sister late Mbuyi Kalenga, with who I shared the birth day, but who unfortunately passed away so soon, rest in peace. May the souls of my aunt late Alphonsine Mbuyi Tshibalanga and my elder brother Benjamin Mbaya Kakasu, rest in peace.

May all members of my extended biological family and friends that I shut the names here, find my deepest gratitude.

I can not forget my best friend ever, late Emile Bantu Mpiana, may his soul rest in peace.

May Emmanuel My Banner family finds my deepest gratitude.

May the workers at Marksbury library in Lexington, Kentucky find my gratitude.

Introduction

The world today continues to live under the threat of global warming, climate change. And if there is one thing the world can agree on, is that the global warming, the climate change is a reality and is a serious threat to the world.

Every two, three, or five years do not pass without us experiencing, or learning through the media, the devastating effects of global warming, climate change. And little by little an evidence begins to emerge: if the day before yesterday it was at the other inhabitant of distant lands, yesterday it was at the neighbor's, perhaps tomorrow it will be at my place.

When we want to know the causes of this phenomenon, some believe that it is inevitable, due to evolution, the aging of the earth. They think that what we are experiencing today is caused only by natural phenomena that occur on earth, and that man can do nothing against it, thus humans can do nothing to stop it.

However, others, including scientists, believe that aside from natural elements, human action would be responsible for the acceleration of global warming, climate change by pouring through greenhouse gases, the CO2, methane gas, nitrous gas, etc in the atmosphere.

Several studies conducted in this regard as we shall see in the coming pages have provided evidence, corroborating human action as the principal cause in the acceleration of global warming, climate change:

A report points out that "because carbon dioxide produced from fossil fuels carries a unique signature, the &13 c (delta c thirteen), scientists can positively identify "human fingerprints and attribute a proportion of observed warming to human activities. Based on that

97% of scientists agree that the global warming, climate change is caused by human activity. These fingerprints include rising levels of carbon dioxide, left in the atmosphere when fossil fuels are burned."[1]

And as we can quote: "over the last two centuries, methane constructions in the atmosphere have more than doubled largely due to human-related activities."[2]

Along the same lines, "an authoritative" 2014 report by the Intergovernmental Panel on climate change, written and reviewed by hundreds of climate experts and scientists from countries around the world, states unambiguously: "human influence on the climate system is clear."[3]

At this level, it must be said that human action is the main cause of the climate phenomenon, although it is not the only cause, because there are natural climate drivers (the energy from the sun reflected back into the atmosphere through terrestrial radiation, the heat from volcanic eruptions, meteor strikes, water vapor, etc) which are among the causes of this phenomenon. And based on available climate data, a correlation can be observed between human activities and the rise in amount of CO2, methane gas in the atmosphere. In this regard, "a line chart showing CO2 over time, that documents dioxide levels since roughly 800,000 years ago, starting around the 1760s shows a dramatic upward spike that is attributed to the industrial revolution."[4]

From what precedes, it appears that human action plays a leading role among the causes of global warming, climate change. And the

---

[0] 2022 U.S. billion-dollar weather and climate disasters in historic context. By ADAM B. SMITH (this is the reference for what is on the back of the cover). PUBLISHED JANUARY 10, 2023

[1] Reports & Multimedia/Explainer How Do We Know Humans Are the Major Cause of Global Warming?
Published Jul 14, 2009 updated Jan 21,2021. Climate Science 101

[2] Importance of Methane/US EPA An official Website of the United States government. Last updated on November 1, 2023.

[3] Reports & Multimedia/Explainer
Op cit.

[4] Idem (Reports & Multimedia/Explainer.)

rise of CO2, among many others in the atmosphere is pointed to as the main element that negatively affects the planet by causing global warming, climate change. And this is a threat to life in the world.

But contrary to that, the CO2's coalition, a new and independent, non profit organization, who speaking of the increase in CO2, invited people to use the old motto of Britain's prestigious Royal society: "nullius in verba: don't take anyone's word for it, see for yourself," is convinced that more amount of carbon dioxide released from the combustion of fossil fuel in the atmosphere will be a net benefit to our planet.[5]

Whereas in 1896 SVANTE ARRHENIUS, "using his knowledge of physics and chemistry, put forward the mathematical relationship between CO2 and atmospheric temperature. His observations led him to theorize - correctly - that human industrial would increase the temperature of Earth's atmosphere, leading to the hypothesis about the relationship between CO2 concentrations and Earth's atmospheric temperature."[6]

Though the CO2's coalition does not agree that higher amount of CO2 in the atmosphere is the main cause of the global warming, they said: "increased CO2 levels have likely produced some warming of the earth and will continue to do so in the future, although with ever decreasing efficiency because of the" logarithmic "dependence of warming on CO2 concentrations."[7]

---

5 J. Tyndall, Heat, A Mode of Motion, Longmans, Green and Company, London, 1875
Carbon Dioxide Benefits the World: see for Yourself. CO2 coalition.

6 DISPOSABLE CITY. Miami's Future on the shore of climate catastrophe.
MARIO ALEJANDRO ARIZA Bold Type Books
116 East 16th street, 8th Floor New York, NY 10003 www.boldtypebooks
Printed in the United States of America
First Edition: July 2020Published by Bold Type Books, an imprint of Persens Books, LLC, a subsidiary of Hachette Book Group, Inc.

7 See, e.g., R.S.J. Tol, The Economic Effects of Climate change, journal of Economic Perspectives, Vol. 23, Number 2, pp 29-51 (2009). Such studies do not fully account for the positive effects of CO2, fertilization and water -

Yet the actual Earth's climate record indicate global warming, climate change. And when you walk in the street, at work, in the family, at school, in our society today, in our small talk, at least eight out of ten people acknowledge that the world is burning, the climate is changing negatively, threateningly. Even the CO2's coalition might acknowledge that it is getting hot, though they say, it is some how getting hot but not the way it is predicted.

But when it comes to tell what is the main cause, though the majority of scientists single out the higher amount of carbon dioxide in the atmosphere, the CO2's coalition without answering explicitly to the question of the main cause of global warming, climate change; thinks that "the negative effects of more carbon dioxide have been exaggerated. And though it might have some negative effects, at the same time more carbon dioxide will have a hugely beneficial effect on agriculture, forests and plant growth in general. The benefits of more CO2 will greatly exceed any harm."[8]

Here is the place to recall, with Jeff Goodell, in his book THE HEAT WILL KILL YOU FIRST, a new facts that have been observed around the world: "In Europe in 2003, 70,000 people died in just few days during a heat wave. In 2021, there were 339 heat related deaths in Maricopa county. In 2022, catastrophic floods in Pakistan displaced 33 million people, which is about 15 percent of the entire Pakistani population. In a 24-hour period, the temperature in downtown Portland jumped from 76 to 114 degrees, the hottest temperature in 147 years. For some insects, these were good times: caterpillars, Maggots, pine bark beetles, the heat goes well with them. Microbes live in 170-degrees hot springs in yellow stone National Park. However all living organisms share one simple fate: if the temperature they are used to - what

efficiency gains. Cited in Carbon Dioxide Benefits the World: See for Yourself. CO2 Coalition.

8 R. McKitrick, HAC - Robust Measurement of the Duration of a Trendless subsample in Global Climate Time serres. Open Journal of Statistics, Vol. 4, pp 527-535 (2014). Dol:10.4236/OJS.214.47050. Cited in Carbon Dioxide Benefits the World: See for Yourself.

scientists sometimes call their Goldilocks zone - rises too far, too fast, they die. And a study in the Lancet, a prestigious medical journal, estimates that 489,000 people die from extreme heat in 2019."[9]

We proceed with the data from the same author: "210 million, increase in number of people facing acute food insecurity. 21% loss in global agricultural production in the last 20 years due to climate heat and drought."[10]

Nowadays, the negative effects caused by climate change are numerous. And the same author notes that "Earth's history is full of wild temperature swings. But for the last three million years or so, while humans evolved, the climate has been relatively stable. The last time the Earth was hotter than it is today was at least 125,000 years ago, long before anything that resembled human civilization appeared."[11]

Referring to negative facts presented above caused by global warming, and regarding the negative climate effects we are witnessing nowadays, it appears that climate change poses a real threat to life on our planet Earth. As you can see after reading the preceding pages, the exceeding amount of carbon dioxide, greenhouse gases in the atmosphere is the main fuel in the global warming, climate change's engine. And it is this thesis that is scientifically observable and documented to these days.

To reduce, decrease or eliminate the amount of greenhouse gases, such as carbon dioxide, methane gas, etc..., in the atmosphere is one of the major keys to slowing down or stopping global warming, climate change.

What to do to reduce, decrease, eliminate the excess greenhouse gases, carbon dioxide, methane gas in the atmosphere? What should be done to address this global threat? Should we surrender to fatalism?

---

[9] THE HEAT WILL KILL YOU FIRST. Life and Death on Scorched Planet. By Jeff Goodell Little Brown and Company. Hachette Book Group 1290 Avenue of the Americas, New York, NY 10104 Littlebrown.com First Edition; July 2023.

[10] Op cit. THE HEAT WILL KILL YOU FIRST.

[11] Idem.

Fortunately, the world in general, especially the scientific community, has not remained indifferent. Several private or state organizations, several scientists individually or collectively after studies, meticulous research, they made proposals for solutions.

But before we come to that, it is important to recall that before the Industrial Revolution, - a period during which humans significantly released enormous amounts of carbon dioxide, methane gas, etc...; resulting in provoking global warming, climate change - ; there were carbon's natural sinks (trees, forests, oceans, etc... to suck the excess greenhouse gases, carbon dioxide, methane gas, etc that could disturb, threaten the equilibrium in the Earth's system.

Then given the current climate threat, it's essential to question about what happened to carbon dioxide natural sinks? Are they overwhelmed, are they no longer functional? Why?

And when talking about the solutions to how to remove the carbon dioxide, methane gas from the atmosphere, we will try to categorize the solutions in two groups, the natural sinks and the non natural or artificial ones.

Returning to solutions for reducing the amount of carbon dioxide, methane gas in the atmosphere, and thus combating climate change, here are some solutions that have been proposed. Francesco Bassetti says: "our oceans are a major carbon sink, absorbing about 22% of the planet's CO2 emissions, thanks to marine plants and animals. Whales among the largest and longest living marine creatures, can not only store large amount of carbon in their bodies, but could indirectly help ocean's carbon intake. Their fecal matter is rich in nutrients that feed the growth of phytoplankton - which take in carbon dioxide from the atmosphere for food. The world's phytoplankton, by one estimate, absorb 10 to 20 billion tonnes of CO2 each year."[12]

---

[12] Whales may be a secret weapon in the fight against climate change. By KRISTIN TOUSSAINT Friday, July 21, 2023 at 5:19 pm.

Furthermore, the same author elaborates by saying: "whales are a key component in ensuring planetary equilibrium."[13] Like many others, here Francesco highlight two elements that can play a major role: "whales can absorb up to 33 tonnes of Co2 on average, while a tree absorbs only up to 22 Kg of Co2 a year. And phytoplankton that are responsible for capturing about 37 billions tonnes of CO2, the same amount as 1.7 trillion trees, or four Amazon forest's worth. And they also contribute as much as 50-85% of the oxygen in the Earth's atmosphere."[14]

We will come back to these two elements when we will elaborate on the reason for this book, even though many solutions have been already proposed. But for now let's continue with solutions within our reach that have been proposed.

The UN-REDD, recognizing that "deforestation account for 17 percent of carbon emissions, advises and gives incentives to countries to preserve their forests as a way to keep carbon dioxide out of the atmosphere."[15]

Amanda Mac Millan said: "in order to avoid the worst effects of climate change, we need to reduce our dependence on fossil fuels and start using clean energy instead. Clean transportation must remain a priority, and regional efforts around the country are helping to boost the electric car."[16]

Getty Wiley too, points out that: "scientists have even conceived to capture and store Co2 waste."[17]

The protection of sea otters as endangered species is important as we can read: "sea otters help fight climate change by acting as guardians of underwater kelp forest that use the Co2 from the atmosphere

---

[13] Why Whales Are Important For Carbon sequestration. Francesco Bassetti August 27, 2020.

[14] The surprising Role of Whales in ocean carbon capture. Friday, Jul 21, 2023 at 5:11 pm.

[15] Protecting Whales to protect the planet.Friday July 21, 2023 at 5:07 pm.

[16] Amanda Mac Millan: Global Warming 101 March 11, 2016.

[17] Sharks and whales are major sink. Getty Wiley/Getty.

to grow. Kelp can absorb between one to ten billion tons of carbon dioxide each year. But kelp forests "are delicious food to urchin. Fortunately sea otters eat urchin, preventing then urchin from eating kelp forests."[18]

Let's proceed with the list of solutions: "The bioenergy with carbon capture and storage, that consist of harvesting plants and storing their carbon,"[19] "stimulating phytoplankton bloom by adding iron to ocean, a form of geoengineering known as iron fertilization."[20] Brad Plumer proposed to pull carbon out of the atmosphere, saying that: "what if we grow trees that sucked carbon dioxide out the atmosphere. Then we burned it, we capture the emissions and pumped them underground? That will be carbon negative."[21]

For Debora Gordon: "we need to manage methane control. The world needs greater methane emission visibility and new methane standards to stop it from wantonly leaking. The R.M.I's Climate Intelligence Program mission is laser- focused on decarbonizing energy supply chains and activating markets for low-and zero-carbon products."[22] In CLEAR CENTER, one can read: "if we are able to reduce methane, such as with dairy digesters, then we can create a cooling."[23]

[18] Human influence-Northern Sea otter.

[19] Want to learn more? In How much carbon dioxide does the Earth naturally absorb? MIT Climate Portal.

[20] Can whales really save the world from global warming? - study - The Jerusalem Post
By Danielle Greyman -KennardPublished June 9, 2023.

[21] How to stop global warming, in 7 steps - VoxBy Brad Plumer/@bradplumer/ brad@vox.com/Nov 2, 2014, 12:51 pm Est.

[22] Methane: A threat to people and Planet - R.M.I July 7, 2021. By Deborah Gordon.

[23] CLEAR CENTER. Clarity and Leadership for Environmental Awareness and Research at U.C Davis Why methane from cattle warms the climate differently than CO2 from fossil fuels/Clear Center. July 07, 2020.

In an official website of the United States government, they talk about "projects to capture and utilize methane."[24] In a UN News too, it is said that: "Cutting methane emissions is the fastest opportunity we have to immediately slow the rate of global warming, even as we decarbonize our energy system."[25]

And "methane-spotting satellite can help us to slow global warming fast."[26] Higher fuel economy and emissions standards combined with increases in incentives for purchasing electric vehicles, are projected to reduce transportation rebated CO2 emissions in 2032 by about 10 percent.[27] "Producing alternative-fuel vehicles, improving the integrity and efficiency of air-conditioning systems and implementing other technologies that improve the operating efficiency of their vehicles."[28]

As for Paul Hawken and his colleagues, they say that as: "the climate change stems from many sources (energy production, agriculture, forestry, cement and chemical, manufacturing, the solutions must arise from the causes of global warming. The needed solutions

[24] Importance of Methane/US EPA an official website of the United States government. Last updated on November 1, 2023.

[25] Climate change: CO2 and methane in our atmosphere reach record levels/UN News 26 October 2022.

[26] Methane: A Crucial opportunity in the climate fight - Environmental Defense Fund. Copyright © 2024.

[27] Environmental protection Agency. Revised 2023 and Later Model year Light-Duty Vehicle Greenhouse Gas Emissions Standards: Regulatory update, EPA-420-F-21-077 (December 2021), https://tinyurl.com/27 caws 5 n (https://tinyurl.cpm/27 caws 5 n) (PDF). Quoted from Emissions of Carbon Dioxide in the Transportation sector. Congressional budget office.

[28] Department of Energy "Electric Vehicle Charging station locations" (accessed on September 2022) https://tinyurl.com/bd8f7rv (https://tinyurl.com/bd8frv). Quoted from Emissions of Carbon Dioxide in the Transportation sector. Congressional Budget Office.

are far more diverse than just those associated with clean energy, and that there are many effective means to address global warming."[29]

In the following lines you will find their proposed solutions:

- "LOWERING NITROUS OXIDE EMISSIONS FROM AGRICULTURE, CUTTING METHANE EMISSIONS from cattle production, and reducing carbon dioxide emissions from deforestation.
- REPOPULATING THE MAMMOTH STEPPE: to keep the planet cool, you want the grasses in sub-polar regions, not trees, and you get grasses when you reintroduce herbivores. Buried in the circumpolar region of the Arctic are 1.4 trillions of carbon, two times more than in all the forests on the planet. When horses, reindeer, musk oxen, and other denizens of frozen north push away the layer of snow and expose the turf underneath, the soil is no longer insulated by its snow cover and is 3 to 4 degrees Fahrenheit colder, a margin of safety the world needs while it transitions away from fossil fuels. If the diverse species of herbivores that once populated the sub-polar region of the Arctic are brought back, permafrost melting can be prevented. And that will be a largest solution."[30]

  We proceed with solutions from the same author.
- "PASTURE CROPPING, increases the use of the land by double-cropping (grains and animals) while reducing impact and increasing carbon sequestration.

  ENHANCED WEATHERING. Natural rock weathering removes approximately 1 billions tons of atmospheric carbon dioxide annually. A conventional pathway to enhanced weathering involves the mining and milling of solicit rocks containing olivine and applying the rock powder to land and

---

[29] DRAWDOWN. THE COMPREHENSIVE PLAN EVER PROPOSED TO REVERSE GLOBAL WARMING. Edited by Paul Hawken. Penguin Books an imprint of Penguin Random House L.L.C

[30] Idem

water so that the soil, oceans and biota can act as "reactors" for accelerated weathering.

-MARINE PERMACULTURE, that may be one of the most extraordinary way to reverse global warming, by reforesting the ocean. Imagine a light weight latticed structure made of interconnecting tubing, submerged 82 feet below sea level, to which kelp can attach.

Marine permaculture supports the functioning of the biological pump so that ocean can do the job they always do."[31]

With the same author: - "INTENSIVE SILVOPASTURE, that sequesters up three times more carbon per acre than grazing alone - ranging from one to four tons per acre in the tropics and averaging 2.4 tons in temperature regions. By adding more cattle, planting different types of trees, and rotating the held more quickly.

-ARTIFICIAL LEAF, perhaps economically viable artificial photosynthesis will not be so artificial after all.

-AUTONOMOUS VEHICLE. The shift to autonomous vehicles rather than private owned cars has to be encouraged.

-SOLID-STATE WAVE ENERGY. The kinetic energy of oceans, which surge with roughly 80.000 terawatt hours of power is extraordinary. A single terawatt is the equivalent of 1 trillion watts and sufficient to provide electricity to 33 million US homes. Because water is nearly one thousand times denser than air, aquaturbines are technically more efficient than wind-turbines."[32]

Still from the same author, - "LIVING BUILDINGS, that produce more energy than they consume and offsetting all embodied carbon as well.

-DIRECT AIR CAPTURE, the idea is to use D.A.C (Direct Air Capture) and carbon dioxide storage to help achieve and maintain drawdown.

---

[31] Ibidem

[32] Ibidem

- HYDROGEN-BORON FUSION, if capture one day's energy from the Sun would power the earth for many years. That is a clean energy that would take the world beyond the age of coal, gas, and oil.
- SMART HIGHWAY, watt way solar road, solar tiles that adhere to existing road ways in order to produce electricity."[33] Still with the same author,
- HYPER LOOP, that is like vacuum tubes which were employed to propel messages, deposits and documents in steel canisters, within buildings and cities. The promise of Hyper loop is speed, its virtue is how little energy it uses to move people and cargo.
- MICROBIAL FARMING,
- INDUSTRIAL HEMP, to be used instead of cotton. Cotton farming requires insecticide, while industrial hemp does not.
- PERENNIAL CROPS, the most effective way to sequester carbon in any agricultural system because they leave the soil intact. Perennial upland rice would minimize deforestation.
- A COW WALK ONTO A BEACH, when cows are fed with some seaweeds, like red Asparagapsus taxiformis, it reduces methane production. Methane, the key waste product generated when cows process the food, drop.
- OCEAN FARMING, Professor Ronald O Singa, has calculated that a global network of sea-vegetable farms totaling seventy thousand square miles, could provide enough protein for the entire world population. Ocean farms require no fresh water, no deforestation and no fertilizer.
- SMART GRIDS, can reduce overall consumption while facilitating the shift away from centralized fossil fuel plants and their greenhouse gas emissions."[34] We proceed with the same author,

---

[33] Ibidem

[34] Ibidem

"BUILDING WITH WOOD, has two benefits: - as they grow trees absorb and sequester carbon which remains stored in timber construction material.

-The process of producing those material generate fewer greenhouse gas emissions than producing wood's alternatives, cement, used in concrete and other building materials.

-RECIPROCITY, when the trees are grouped together, are benefiting one another in some way. The "wood wide web" as Simard calls it, is an underground intern through which water, carbon, nitrogen, phosphorus, and defense compounds are exchanged."[35]

After having set out the different valuable solutions proposed by eminent scientists or organizations of great expertise, let's come back to say something about the reason for our book. Why am I still writing about a proposal to fight against global warming, climate change? Here it should be noted that in what has already been proposed, there are relevant points not to be overlooked.

But in our book we want to start from what already exists and see what can be added to make the result, the removal of the CO2 , methane gas's excess in atmosphere more effective.

Considering the existing options to fight global warming, climate change, certain solutions will be avoided because of drawbacks, such as high economic costs and the considerable financial investment's needs for their realization, either because of their feasibility which is still uncertain; or because of the required technology which is not yet realistic.

Like a bee that uses elements that already exist in nature, to produce honey (something new that did not exist before), we have used the existing literature on this topic (Fighting global warming, climate change), to produce a solution with new element capable of strengthening the fight against global warming, climate change.

[35] Ibidem

Speaking of what have been already proposed and their limits, we think for example to:

- bio-energy with carbon capture and storage (B.E.C.C.S) that can remove more CO2 from the atmosphere. However for this to be effective it will needs a lot of land, which land might be needed for farming, housing, factories etc...And the technology to be used is not only expensive, but also uncertain.

  The Intergovernmental Panel on Climate Change (I.P.C.C), the world's authoritative body on global warming, climate change's matters, estimates that "the technology to capture carbon emissions from coal plants and bury it underground is still in its infancy. But if that technology proves unworkable or limited, the task of cutting emissions becomes twice expensive."[36]

-The Solid-State Energy, is one of the valuable solutions, but its affordability may be a hindrance to its realization, when we look at what it can cost. And that is without taking in accounts the costs that will be related to "maintenance, replacement parts, servicing in high seas and under water cables to transfer the power."[37]

-The Hydrogen-Boron Fusion, that can give a clean virtually source of energy, and take the world beyond the age of coal, gas and oil. However to these days it is not yet a reality, though the California-based T.A.E Technologies "plans to build the

---

[36] How to stop global warming, in 7 steps-Vox By Brad Plumer/@bradplumer/ brad@Vox.com/ Nov 2, 2014 12:51 pm

[37] DRAWDOWN. THE MOST COMPREHENSIVE PLAN EVER PROPOSED TO REVERSE GLOBAL WARMING
Edited by Paul Hawken Penguin Books.
An imprint of Penguin Random House L.L.C 375 Hudson Street New York, New York 10014
Copyright © 2017 by Project Drawdown.

world's first prototype hydrogen-boron fusion power plant in the early 2030s."[38]

-The Hyper Loop, that can be helpful with its high speed, but its safety and what it can cost would be the obstacles to its realization.

In the lines above, we talked about some solutions carrying hope, but which have high economic costs, either because they require a lot of space, while space is nonetheless in competition for other needs, or its technologies are still uncertain. All this can be obstacles for their achievements.

However, couldn't alternatives - that will suffer less from exorbitant economic costs, not requiring more space, that is in competitions for other needs - be found? Alternatives that may benefit from remedy that may have already been tested and used in a different field, for a different purpose than the purpose of this book, but which will be rational, efficient and effective in the fight against global warming, climate change?

This is where comes our contribution, to be presented to the world in general to fight against global warming, climate change. This proposal is addressed to the world in a general way, and to political decision-makers, to leaders of organizations around the world, including whoever who can help to make it a reality, from words to action. It is a strict and strong measure, which if applied,will be a protection against the extinction of whales and will favor the increase of its population. Thus boosting the resilience of the climate's system. In the pages above and going through the book, it is shown that whales and Sea otters play a major role in removing or reducing the excess of CO2, methane gas from the atmosphere.

Fortunately these two animal species are aquatic, live in water (oceans); and as we know the body of water covers two-third of the Earth's space. With this, first of all the problem of space will be solved

[38] HOT FUSION: EXASCALE COPUTING DRIVES EARTHLY EFFORTS TO PRODUCE THE ENERGY THAT POWERS THE SUN. 01/08/2024 Steve Koppes.

as they are aquatic animal species. Secondly, as we can read from the existing literature presented in the preceding pages, it has been elaborated how whales and sea otters are a major key in the fight against global warming, climate change. And that is more than necessary to stop the extinction of these two animal species.

In this regard, several attempts have been made, but with timid results or unconvincing because to these days; the threats to these two species are still relevant.

That is why we propose, as you will read in chapter 5, a rigorous and effective measure to stop the threat of extinction, and promote the increase of the population of these two species.

METHOD

We want to talk here about the method used in the production of this book. But before that, let's talk a little about the explanatory element that determines social life. In this regard, sociologists' view are divided. Without minimizing the intermediate point of view, there are two extremes:

- those who state that social life can be explained by social consciousness, or more specifically the culture. And
- those for whom social existence is the driving force of social life. In this regard, I believe that social existence mainly influences social life. In fact social existence which is expressed within a social sphere, this social sphere being the environment in which man lives: the family, clan, tribes, ethnic groups, nation. It is heavily influenced by how one fits into society and is treated by it.

The viewpoint of the proponents of social existence as the primary element determining social life is summarized in what Karl Marx said: "it is not consciousness that determines life, but life that determines consciousness. The material conditions of social existence determine individual consciousness. The production of ideas, representations and consciousness is first directly and intimately interwoven with material activity, and in the material commerce of men, it is the language of real life."[39] And as we know the object of Sociology is from the outset society considered as an organized totality.

But given that social existence and society are threatened by global warming, shouldn't sociology engage in a kind of "outside curiosity," external perspective in order to find the solutions to fight or stop global warming, climate change on Earth, though it is generally the subject of the sciences other than Sociology?

In fact global warming, climate change are fundamentally linked to social factors (societal structures and human actions contribute to global warming, climate change and shape their effects).

It can be noted here that although, when speaking of devastating effects of global warming, climate change; one thinks of physical destruction, such as the physical destruction of infrastructures like roads, railways, airports, entire cities, loss of life, interference in the economy, in the politics etc.... Furthermore, global warming, climate change affect the social fabric, the social's life, the social relationship, the society which is itself the very object of Sociology.

Considering the Earth as a system composed of interacting parts, each with a function for the maintenance of the system, the structural-functional method will be used in this work, book. This method is based on the notion of a system composed of interacting parts in their function of maintaining the system, and the dysfunction of one of the parts influences the functioning of the whole system.

In the following pages, you will be able to see how this method is used.

---

[39] W Connection Philo Blog
Le blog philo de Herve MOINE, professor of philosophy Itinerant-laphilo.com c Rechercher dans le philo Blog.

Chapter 1

Some Information About The Earth's Planet As A System

After the introduction, here is the first chapter in which some information about planet Earth will be provided. This is without claiming to provide more elaborate information about planet Earth, because the goal is simply to allow the reader to get an idea of what planet Earth is. This will allow the reader to follow along with us as we progress. That is why we named this chapter: Some information about the Earth's planet as a system.

We are talking about global warming, climate change which are phenomena that happen on planet Earth. It is important that we talk a little about the planet Earth where we live, and where global warming, climate change are happening.

The Earth's planet as a system is one of the components of a system called solar system. The solar system as we know from geography, is made up of the Sun, the eight planets moving around it, the stars, moons and other objects.

"Third planet ranging from the Sun, the Earth has a particularity as it offers conditions for life when we compare it to other planets. For example Venus is the "hottest planet with a temperature of 867 degree Fahrenheit due to it's thick atmosphere with higher amount of carbon dioxide which traps heat inside it, and clouds made up of sulfuric acid."[40]

The Earth's planet is made up of five subsystems which are in interaction:

- "Geo-sphere (lithosphere): all the rocks and "hard parts" of the Earth
- Hydrosphere: all of Earth's water
- Biosphere: all living organisms
- Atmosphere: the blanket of gases surrounding the planet.
- Cosmosphere: the universe surrounding the Earth."[41]

The five subsystems interact each other to maintain certain conditions on Earth. Here are some examples of interactions: "the evaporation that occurs in the hydrosphere forms the medium for cloud and rain formation in the atmosphere. And the atmosphere brings back this water to the hydrosphere in the form of rain. The Hydrosphere provides the necessary moisture required to weather and erode rocks in the Geo-sphere. The Geo sphere in turn allows the ice to melt and the water bodies to flow back into the oceans. The atmosphere provides the required heat and energy for the breakdown and erosion of rock in the Geo sphere. The Geo sphere in turn, reflects the sun's energy to the atmosphere. And it is all the interactions that makes life as we know it.[42]

While knowing that the five subsystems are in interaction, our main focus will be on the atmosphere. This choice is dictated by the fact that since we are talking about global warming, climate change

[40] Solar system 101/National Geographic.

[41] Earth's systems.
Related resources: National Geographic
Headquarters 1145 17th Street N.W Washington, DC 20036. Society is a 501 © (3) organization. © 1996-2024 National Geographic society. All rights reserved.

[42] Fundamentals of Earth's Sphere Interaction between Spheres.

whose main cause is the excess of CO2, methane gas, etc... in atmosphere, it is relevant to focus on the atmosphere.

A layer of gas, the atmosphere is made up of "Nitrogen (N2) 78% of air, Oxygen (O2) 21%, the inert gas argon (Ar) 0.93% and trace amounts of Carbon dioxide (CO2), Neon (Ne), helium (He), Nitrous oxide (NO), Xenon (Xe), Ozone (03), Iodine (I_2), Carbon monoxide (CO), and Amonia (NH_3)."[43]

Being a subsystem of the Earth's planet, the atmosphere is a system too, with its components in interaction. It is important to note here that "carbon dioxide naturally composes about 0.03% of the atmosphere."[44]

From what is said above, it appears that naturally the nitrogen and oxygen make up 99% of the volume in the atmosphere. Looking at their proportions in the atmosphere, some people might be tempted to think that the other gases in atmosphere are useless. Contrary to that, we should understand that the other gases too play each, a specific role, function to contribute to the conditions for the Earth's planet to bear life as we see it.

And the equilibrium between different gases composing the atmosphere, based on their natural proportions, plays a major role to sustain pro-life's conditions in our home planet, Earth's planet. For example, the two natural most abundant gases: nitrogen and oxygen would not play the roles filled by the other gases like carbon dioxide, though naturally in minor amount proportionally in the atmosphere. Living organisms, plants, animals, bacteria, etc... use carbon dioxide to be alive. For example plants need carbon dioxide for photosynthesis.

Though carbon dioxide is one of the heat-trapping greenhouse gases, it has a role to play in Earth's planet, our home planet. So the carbon dioxide gas is needed on Earth's planet, but in the amount that does not upset, affect the natural equilibrium, without which the car-

---

[43] U.C.A.R/Center for Science Education S.C.I.E.D
What's in the Air/Center for Science Education.

[44] Idem

bon dioxide, instead of being beneficial to the system, will be dangerous, destructive to the system.

By reading the history of the Earth, it is possible to see that originally, there was an equilibrium between the different gases that make up the atmosphere. But it is over time that the change in the quantities of gases in the atmosphere began to be observed. When this happened, it started causing changes in the climate's indicators such as the temperature, precipitation, seasonal disruptions, etc...

And speaking of temperature, which is one of the indicators of global warming, climate change, we can read: "global temperatures on Earth have increased by at least 1.1 degree Celsius (1.9 Fahrenheit) since 1880, according to ongoing temperature analysis led by scientists at NASA's Goddard Institute for space studies (G.I.S.S).[45]

We conclude this chapter by asking the question, what caused or brought about the rise in CO2, methane gas, nitrous oxide in the atmosphere?

In the following chapter we will have the answer to this question.

[45] Sharad Chouhan 12th in Geography Major & Sanskrit (language). BA FIRST YEAR. 1y Cited in Does the average global temperature ever decrease? - Quora.

Chapter 2

Human Action and its Consequences in The Global Warming, Climate Change

We begin this chapter with Jeff Goodell's say: "Earth's history is full of wild temperature swings. But for the last three million years or so, while humans evolved, the climate has been relatively stable. The last time the Earth was hotter than it is today was at least 125,000 years ago, long before anything that resembled human civilization appeared."[46]

Today, global warming or climate change is real and when talking about global warming or climate change's effects, we can not quote two, three or four of them without referring to the high rise of temperature for which the effects are no longer debatable. Speaking of temperature rise, it is important to note that even without human interference, temperature rise can occur, caused by natural causes "such as changes in solar radiation or volcanic activities that are estimate to have contributed less than plus or minus 0.1 degree Celsius total warming between 1890 and 2010,"[47] and others like natural

[46] THE HEAT WILL KILL YOU FIRST. Life and Death on a Scorched Planet. By Jeff Goodell Little Brown and Company. Hachette Book Group 1290 Avenue of the Americas, New York, NY 10104 Little Brown.com First Edition: July 2023

[47] Data Viz showing how 97 percent of scientists agree on climate. In Reports & Multimedia/Explainer. How do we know that humans are the major cause of

emitting termite mounds, CO2 uptake by plants, etc... "All of these factors have profoundly shaped the Earth's climate. However none of them sufficiently explain the recent dramatic increase in global temperatures. For that, scientists need to include human influences."[48]

The planetary threat that is the subject of this book has been caused by the fact that in addition to natural causes there is, on the one hand - to find solutions to our different needs - we have used among solutions, primarily the sources of energy that emit greenhouse gases which are nonetheless the causes, the main drivers of global warming, climate change. And on the other hand, the flora, fauna, hydrography etc..., which constituted the natural sinks; have been destroyed or significantly reduced. And thereby reducing the capacity of natural sinks to absorb the excess of CO2, methane gas, etc... in atmosphere.

This is how little by little perhaps without realizing it, man has overloaded the atmosphere with pollutants: the carbon dioxide, methane, nitrous oxide, various chlorofluorocarbons, etc... that are occupying the atmosphere more and more. They form a kind of blanket, which let penetrate solar energy, solar rays toward the Earth's surface, but preventing them from escaping. These energies become compressed between the Earth's surface and the cover formed by the effects of greenhouse gases, caused by human activity. This is how over the years - accelerated by human action - the quantity, the concentration of CO2, methane gas, nitrous oxide, and other pollutants have increased in the atmosphere. For example, the CO2 is the main contributor to the global warming, climate change, exerting a major influence to them with effects lasting longer.

Everything must have started somewhere. According to the existing literature on the history of the increase in CO2, there is a correlation between the CO2, methane gas, nitrous oxide's increase and the Industrial Revolution. "Carbon dioxide concentration have increased substantially since the beginning of the industrial era, rising from an annual average of 280 ppm in the late 1700s to 414 ppm in 2021 (aver-

Global Warming? Published Jul 14, 2009 updated Jan 21, 2021.

[48] Idem

age of five sites) - a 48 percent increase. Almost all of this increase is due to human activities. The concentration of methane in the atmosphere has more than doubled since preindustrial times. The increase is predominantly due to agriculture and fossil fuel use.

Over the past 800,000 years, concentrations of nitrous oxide in the atmosphere rarely exceeds 280 ppb (parts per billion). Levels have risen since the 1920s, however, reaching a new high of 334 ppb in 2021. This increase is primarily due to agriculture."[49]

We just talk about the concentration of carbon dioxide, methane and nitrous oxide in the atmosphere, "the three largest greenhouse gases."[50]

However, in addition to these three, there are also many others such as the Fluoronated gases, Hydrofluoride etc... which also contribute according to their proportions, to the discharge, to the excess of gases in the atmosphere.

We said that over concentration, or excess of these gases, because each gas has an amount, a proportion that does not disrupt the natural equilibrium, that does not harm the natural balance between different gases that make up the atmosphere, different parts that compose the planet. But in case this balance is threatened, the planet was equipped with natural sinks to absorb the excess gases that threaten the atmospheric balance.

In chapter 1, it was stated that the planet Earth is like a system composed of different interacting parts. The atmosphere being a subset of the planet, it can also be compared to a system composed of different parts. And it so happens that when one of the parts malfunctions, or is in overabundance compared to its initial proportion and brakes the natural balance, it affects the rest of the system and the whole system.

---

[49] Climate change Indicators: Atmospheric Concentrations of Greenhouse Gases/ US EPA

[50] Climate change: CO2 and methane in our atmosphere reach record levels/UN News
26 October 2022.

And you realized that in the introduction, we announced that we will use the structural-functional method, a method based on the notion of a system.

This is how the system endowed with the favorable elements for maintaining the necessary balance for the survival of the entire system, for the well-being of humanity, of the society, the planet. And during the normal course of the system, while equilibrium is present, the society enjoys this balance because its life depends on it. And there, the society suspects nothing regarding actions that can harm this equilibrium, especially if this society is not informed. That is how we take everything for granted, without knowing that there are elements contributing to that equilibrium, on which the life of the society, the planet, the world depends.

Thus, since a certain point in history, specifically the Industrial Revolution, this equilibrium began to be threatened, until it was broken, giving way to what we are experiencing today, the global warming, the climate change. What happened?

In the ancient times, people according to their culture used things like wicks soaked in palm oil and animal fats, which they carefully diluted, to light their way at night. But with the Industrial Revolution, people began to use electrical energy and other types of energy emitting greenhouse gases for light, and especially for different needs: producing foods, using electricity and other sources of energy to produce in large quantities consumer products, equipment, the needs of dressing, building homes, manufacturing for various uses.

But to produce electricity, we need to use and still use mainly the fossil's fuel source of energy. "according to recent data, over 60% of the world electricity is currently produced mainly by fossil fuels."[51] And "around 82% of the world's energy mix is currently derived from fossil fuels, meaning the vast majority of factories and industries worldwide are primarily powered by fossil fuels."[52]

---

51 Energy Institute releases 2024 Statistical Review of the World. Jun 21, 2024.

52 What is the Percentage of factories and industries worldwide, powered primarily by energy from fossil fuels compare to those...

And we know the link between fossil fuels, carbon dioxide, global warming, climate change. But to produce foods, equipment, etc., we use energy to run factories. Electrical energy has since been used in manufacturing companies, offices, homes, entertainment venues etc. To bring products from manufacturing sites, farms and production businesses, people needed to transport these goods. They needed means of transport to move their goods and themselves from one place to another. For this, they created networks of roads, rail, and air communications.

This has created problems to the environment. In producing electricity and ensuring the transportation of people and their goods, people had not yet realized how much the atmosphere will be overloaded with greenhouse gases.

As for electricity production, it is still a minor part that is produced by using wind energy, solar energy, or some other renewable sources which emit less pollutants in the atmosphere. Although there are hydroelectric dams that emit few greenhouse gases that causes global warming, climate change, most electricity is still generated by burning oil, coal, gas; all emitting carbon dioxide.

As an example, "in the United States, the burning of fossil fuels to make electricity is the largest source of heat-trapping pollution, producing about two billion tons of CO2 every year. Coal-burning power plants are by far the biggest polluters."[53]

And according to FOSSIL FUELS/E.E.S.I: "fossil fuels including coal, oil and gas have been powering economies for over 150 years and currently supply about 80 percent..."[54] in the same path, "according to recent data, around 82% of the world's energy mix is currently derived from fossil fuels, meaning the vast majority of factories and

[53] Amanda Mac Millan: Global Warming 101
March 11,2016.

[54] FOSSIL FUELS/EESI
Jul 22, 2021

industries worldwide are primarily powered by fossil fuels like coal, oil, and natural gas."[55]

Electricity's production put aside, thousands of people move around, transport goods, using means of transportation by roads, rail, waterway, maritime, and air. By using these means, devices, equipment that use gasoline or other type of fuels to propel their engines, these machines, dump million tons of CO2 in the atmosphere. For example, the US's second largest source of carbon pollution is the transportation sector, which generates about 1.7 billion tons of CO2 a year."[56]

We can realize that the threat is real when we consider the number of all the means of transportation in the world. And "almost all 95% of the world's transportation energy comes from petroleum based fuels, largely gasoline and diesel."[57]

We talk about transport communication networks. The construction of airports, river ports, maritime ports, railways, roads, bus stations, train stations, etc.., to be built, need spaces. But those spaces might be the spaces previously occupied by the flora, fauna, hydrography, etc.., which were nevertheless the CO2 natural sinks; playing a significant role in maintaining ecological balance for the well being of people, of the society. But unfortunately, all of that have been destroyed or significantly reduced, to make way for communication networks to meet the transportation needs of people and their goods.

In addition to these communication networks, people need space to build their farms, factories, homes, hospitals, schools, universities, libraries, sport fields, entertainment's grounds, etc... But on these spaces, as mentioned in the preceding paragraphs, those spaces were often occupied by natural sinks, that contribute to the natural balance, thus contributing to the well being of the planet, of the people, the society.

---

[55] Ibidem (What is the percentage of factories and industries worldwide, powered primarily by energy from fossil fuels compare to those relying...)

[56] Amanda Mac Millan:Global warming 101March 11, 2016

[57] Global Greenhouse Gas Overview/US EPA September 10, 2024

That does not mean that before industrialization happened, no space was occupied, no flora, fauna or hydrography, was destroyed. But the extent of the occupation of space - accelerated by the increase in population and the desire to produce in large quantities, more and more - this occupation of space and the increased emission of polluting elements into the atmosphere differ from those before the Industrial Revolution.

We said that people occupied the space, among other things to build their farms. To clear the land, to get the spaces for farming, trees have been cut down, acres of forests have been destroyed. And on top of that, when farming livestock, it does not go without negative influence on environment, increasing global warming, climate change.

Herds of cattle being farmed in large number, drop their excreted fecal matter that makes piles of wastes producing bio genic methane, a "second most abundant anthropogenic GHG after carbon dioxide"[58] and "Methane is responsible for around 30% of the rise in global temperatures since the industrial revolution,... The concentration of methane in the atmosphere is currently around two- and - a half times greater it's pre-industrial levels. Estimates of methane emissions are subject to a high degree of uncertainty, but the most recent comprehensive assessment provided in the Global Methane Budget - Suggest that annual global methane emissions are around 580 Mt (megatons). They includes emissions from natural sources (around 40%) and the remaining 60% which originate from human activity (anthropogenic emissions). The largest anthropogenic source is agriculture, responsible for around one quarter of emissions."[59]

If we combine global grazing land with the amount of cropland used for animal feed, livestock accounts for 80% of agricultural land

---

[58] Importance of Methane/US EPA An official Website of the United States government
Last updated on November 1, 2023

[59] Methane and Climate Change - Global Methane Tracker 2022 - Analysis - IEA

use."[60] And in livestock farming, machines, equipment that use fossil fuels are used and release carbon dioxide emissions.

When farming crop, farmers - targeting to maximize their productions - use pesticides and fertilizers that can contain nitrogen and other harmful pollutants to environment. And some research indicates that a large amount of methane is coming from "biogenic sources such as wetlands and rice paddies."[61]

We know that crop farming too uses tractors, equipment that use fossil fuels, thus causing carbon dioxide emissions. Before we move from talking about agriculture, let say with HANNAH that, "if we rewind 1000 years, it is estimated that only 4 million square kilometers - less than 4% of the world ice - and desert - free land was used for farming. But nowadays, almost half (44%) of the world's habitable land is used for agriculture. In total, it is an area of 48 million square kilometers (km2). That is around five times the size of the United States."[62]

We talk about fossil fuels being used to run factories producing foods, consumer goods, capital goods, etc... Let talk about factories using wood as their raw material, timber harvesting, etc... to make papers, electrical poles, different consumer goods made of wood, etc... In addition, these factories, companies, etc... use sources of energy from fossil fuels to be run, they cut down trees, doing deforestation for huge spaces of forests, to sell timbers, logs and different wooden products. And we know that deforestation is one of the causes of the high rising carbon emissions, because forests help to sequester carbon dioxide, but when they are cut down, they release carbon dioxide they have been absorbing.

---

60 HOW MANY ANIMALS ARE FACTORY FARMED? The majority of farm animals in the world are factory - farmed

61 Climate change: CO2 and methane in our atmosphere reach record levels/UN News
26 October 2022

62 Ibidem, HOW MANY ANIMALS ARE FACTORY FARMED? The majority of farm animals in the world are factory-farmed. By HANNAH RITCHIE September 25, 2023

Worldwide trees are being cut down though there are warnings or restrictions. Woods are needed for different uses to make products of various uses, for building, automobile industries, electrical poles, etc... And there we are talking about deforestation, a global threat that deserves real attention.

However, it must be recognized that efforts have been and continue to be made through the world, in different regions of the world to fight against deforestation. For example, from 1990 to 2000, in ten years, Africa lost over 12 millions acres of their natural forests. Only Europe managed to increase its forest cover. According to these data, Africa was the first, South America the second, Asia the third, North America the fourth and Oceania the fifth to have registered loss in this matter."[63]

But from 2013-2023, it was no longer Africa that was in first position. "From 2013-2023, the continent with the highest deforestation rates are South America (particularly Brazil), Africa, and South Asia with the most significant losses occurring in the Amazon rainforest, while regions like Europe and North America have generally seen a decrease in deforestation rates or even slight forest regrowth, however, recent data shows a concerning increase in tree cover loss globally.[64]

Here it should be noted that according to a report "almost 16 million acres of forest were razed in 2023."[65] Yet in 2020 in their meeting, the New York Declaration on Forests (N.Y.D.F) - a renowned framework that calls for global action to protect and restore forests - created and endorsed in 2014 at the United Nations Climate Summit, around 140 countries concerned about putting an end to deforestation, made a vow and set up measurable progress to be reached by 2030.

---

[63] Published by Facts on File, in. Cited by Honore Matamba, in PROPOSAL OF SOLUTION ON THE CHAOTIC POLITICAL SITUATION IN DRC Copyright © 2011 by HONORE MATAMBA www.xlibrispublishing.co.uk

[64] Deforestation "roaring baca" despite 140 - country - vow to end destruction. (Deforestation) The guardian. This article is more than 2 months old Damian Carrington Environment editoTuesday 8 October 2024 00:01 EDT

[65] Other countries that made progress towards the 2030

Among the N.Y.D.F's ten goals, is to "halt natural forest loss by 2030, restoring 350 million hectares of degraded landscapes and forest lands."[66] Already in 2022, the Forest Declaration Assessment - an independent civil society that manages to track progress made to reach the N.YD.F's targets -, reported that the world was 21 percent short of achieving its yearly goal. Total loss was four percent higher than in 2021."[67]

In 2023 too, an analysis shows that "the destruction of global forests increased and is higher than when 140 countries promised three years ago to halt deforestation by the end of the decade."[68]

In the same pathway, Forest Declaration Assessment report reveals that "available evidence shows that restoration progress is likely nowhere near the need."[69]

And to the question, is deforestation still a threat in 2024? The answer is yes according to the Forest Declaration Assessment: "deforestation remains a significant threat in 2024, although the rate of deforestation has slowed down in some regions; however persist regarding the ongoing destruction of forests due to the factors like climate change, agriculture expansion, and unsustainable logging practices, particularly in area like the Amazon rain forest where deforestation is still a major issue despite recent declines in certain regions."[70]

In the light of the above, deforestation still a threat up to now in 2024. And we know that deforestation has two main causes: natural and human related. Natural when caused by forest fires, natural disas-

---

[66] Forest Declaration Assessment. What is the New York Declaration on Forests?

[67] (deforestation facts to know in 2024 (plus solutions) /fsc.org September 19, 2024

[68] Ibidem. Deforestation "roaring baca" despite 140 - country vow to end destruction. (Deforestation) The Guardian. This article is more than 2 months old.

[69] Forest Declaration Assessment Transition Risk: the 2024 Assessment reveals risks and opportunities for forest protection. Assessment Report/October 8, 2024

[70] Is deforestation still a threat in 2024? - Google search.

ters, drought, plant diseases etc... Human related when caused by agricultural activities, and - "the world wildlife fund announced that 80 percent of global deforestation is linked to altering natural landscapes for crops and livestock"[71] - , logging, urbanization, mining etc...

With what is said in the paragraph above, it is obvious that 80 percent of the global deforestation causes are human related. Thus human related causes are the main causes of deforestation, as the world is transforming the natural landscapes to crop and livestock's farming, clearing lands for housing, building roads, mining, etc...

As if this is not enough, man in his activities, as we already said in the pages above, releases enormous quantities, significant amount of CO2, methane, nitrous oxide in the atmosphere. The excess of these greenhouse gas leads to global warming, climate change. Global warming or climate change leads to drought that is in co-relation with wildfire which is among the drivers of the tree cover loss worldwide, thus deforestation. With deforestation, there is more CO2's emissions through the release of the CO2 kept for years in the live trees. And with more CO2's release, the global warming or climate change is being fueled.

And all this started somewhere: in biosphere - one of the spheres, parts composing the planet Earth - , the forest cover is affected due to human activities. The planet Earth as a system is composed by different spheres in interaction. As the biosphere is affected by the deforestation, and being in interaction with other spheres, it affects the atmosphere, which is the sphere made up of gases. In fact deforestation causes the release of CO2 in atmosphere. When we are cutting down trees, when destroying the forest cover, the CO2 that was stored in the trees, in the vegetation is released off in the atmosphere, by then polluting.

The CO2 gas released by deforestation adds to the CO2 already in atmosphere caused by other human activities like factories, mining, air, road, sea and railway transportation's means etc... make the

---

[71] 9 deforestation facts to know in 2024 (plus solutions)/fsc.org
The world lost 6.6 ha of forest in 2022

CO_2' s amount in atmosphere to exceed the limit, amount of CO_2 needed in the atmosphere, thus breaking up the natural balance of gases in atmosphere.

Deforestation releases the CO_2, and more CO_2 causes the rise of temperatures. The rise of temperatures causes permafrost thawing, thus affects the hydrosphere, the sphere of Earth that is made up of all of the Earth's water. In addition to that permafrost leads to more CO_2 emission as the dead materials in permafrost release CO_2 that was kept in them.

The same way, deforestation releases CO_2 and more CO_2 leads to temperature's rises, leading to global warming, climate change, that leads to drought, wildfires. Drought, wildfire cause forest cover loss, deforestation and leads to CO_2's emission. More CO_2's emission leads to temperature rise, thus to global warming. In the same way, deforestation causes CO_2 release. More CO_2 causes the rise of temperature, the rise of temperatures, the rise leads to the rise of sea level, thus floods affecting cities, roads, human life. Floods cause sometimes dangerous disruptions of the environment, affecting life of people in the area of floods and around. It affects negatively the livelihood of societies, provoking forced migrations, sweeping cities etc. It can even causes forest cover loss, that leads to global warming, climate change.

Let's talk a little bit about flood, an overflow of water that submerges a land that was usually dry. The flood is a natural disaster generally caused by heavy rainfall, storm surge, huge snow melt and tsunami. It is relevant to say that the three first causes of flood (heavy rainfall, storm surge, and huge snow melt) have something in common as their direct or indirect potential trigger. That is the rise of temperature, and we know the link between the rise of CO_2 and the rise of temperature in atmosphere. Thus it is obvious when Gregory J. Retallack says: "a direct consequence of rising CO_2 is increasing devastating flooding."[72]

---

[72] GSA TODAY
Article, pp 4-8/Abstract/PDF Flooding Induced by Rising Atmospheric Carbon Dioxide

Since the rising or decreasing of CO2 influences directly or indirectly the flooding - topic in this part of our book - it is relevant to look just for our information a little bit, at:

1. The CO2 emission historical records starting from or around the commercial whaling era.
2. The deforestation historical records starting from or around the commercial whaling era.
3. The whaling history's records.

We look at the two last: deforestation and commercial whaling, because the two being natural CO2's sink, their influence on the amount of carbon dioxide in atmosphere is vital. This is because if we stop the monstrous deforestation, there will be a positive impact, as more forest cover plays a major role in reducing the amount of carbon dioxide in the atmosphere.

In the same way, if we stop the monstrous commercial whaling, there will be a positive impact, as more whales will play a major role in reducing the amount of carbon dioxide in atmosphere.

That is why in the following lines we will give, in one hand some information on CO2 emissions 1750 - 2021. And in the other hand, we will give some information on forest's loss. We need all this information around whaling era, because whaling itself has an influence on carbon dioxide's amount in the atmosphere.

Gregory J. Retallack*
Dept of Earth Science, University of Oregon, Eugene, Oregon 97403 - 1272, USA

CARBON DIOXIDE EMISSIONS

Based on the graph of "atmospheric carbon dioxide amounts and annual emissions (1750 - 2021)[73]

From 1750, the carbon dioxide emission in atmosphere's line is flat, being around zero gigatons. It is from around 1850, that the line started ascending - showing increase of CO2's emission - going from around zero in 1850 to around 34 billion metric tons in 2020. In other words, from 1850, this line keeps going up, accumulating since around 1850. This means that the amount of Co2 in atmosphere keeps increasing since 1850.

When it comes to Earth forest's loss: "in 1700, the forest was 52% of the total land.

In 1800, the forest was 50%.

In 1900, it was 48%.

In 1950, 44%.

In 2018, 38%"[74]

From the data above, it appears that for the forest's loss too, the line goes up, meaning that the reduction of the forest cover is increasing up.

Let's look at 30 years of Deforestation and Forest Growth.[75]

---

[73] Climate change: Atmospheric Carbon Dioxide. (https://www.climate.gov/media/14566)
By REBECCA LINDSEY/REVIEWED BY EDDLUGOKENCKY
Published June 23, 2022

[74] Deforestation has caused the loss of ⅓ of the World's forests/World Economic Forum Nature and Bio-diversity. Here's how the Earth's forest have changed since the last ice age.
April 7, 2022 Our World in Data.

[75] (Mapped: 30 years of Deforestation and Forest Growth, by Country. Get visual capitalist's latest publication: 2025 Global Forecast Report.
Published 3 years ago on December 29, 2021
By NICCOLO CONTE Article/Editing: Niccolo Conte

30 YEARS OF DEFORESTATION AND FOREST GROWTH

From this data, we have:
- Africa with -14.29%
- North and Central America -0.34%
- South America/Caribbean -13.30%
- Europe 2.34%
- Asia 6.64%
- Oceania 0.00%

The following tables (table 1 and table 2) will help to give more view about some information related to this section.

Here is table 1.

8 Most catastrophic floods. (1)

Name	year
Johnstown flood	May 31 1889
Central China Flood	Summer 1931
Great Drawning	January 1362
Few Floods IN	1841
The US most iconic	Spring 1927
Italy cultural thred	1966
Blamed on mother	June 1938
Historians...	January 30 1607

Global land and ocean May temperature anomalies. (2)

Date	Anomaly in degree celsius	Rank
May 1889	-0.26	24
May-31	-0.13	63
May 1927	-0.2	38
May 1966	-0.11	71
May 1938	-0.03	92

Ten warmest years 1880-2022. (3)

Year	Anomaly in degree celsius	Anomaly in fahrenheit	Rank
2016	0.99	1.78	1
2020	0.98	1.76	2
2019	0.94	1.69	3
2015	0.93	1.67	4
2017	0.91	1.64	5
2022	0.86	1.55	6
2021	0.84	1.51	7
2018	0.82	1.48	8
2014	0.74	1.33	9
2010	0.72	1.3	10

Global mean annual temperature average per decade. (4)

Decade	Degree celsius	Fahrenheit
1880S	13.73	56.71
1890S	13.75	56.74
1900s	13.74	56.73
1910s	13.72	56.7
1920s	13.83	56.89
1930s	13.96	57.12
1940s	14.04	57.26
1950s	13.98	57.16
1960s	13.99	57.18
1970s	14	57.2
1980s	14.18	57.52
1990s	14.31	57.76
2000s	14.51	58.12

Sources:

1. 8 of History's Most Catastrophic Floods.
2. Global time series/climate at a Glance/National Centers for Environment (NCEI)
   June U.S Release: Tuesday, 11 July 2023, 11:00 am EDT Global Data Time scale: 1 - MonthMonth: May start year: 1850 End year: 2023 Region: Global
3. N.C.E.I monitoring reports by National Hydro-meteorological Service (N.H.H.Ss ; peers of the U.S National Weather Service) and W.M.O's 2022 provisional statement. Cited by Annual 2022 Global Climate Report/National Centers for Environmental Information Ten warmest years (1880 - 2022).
4. Annual Average Temperature History for Earth - current Results. Weather and Science facts. History of changes in the Earth's Temperature.

By LIZ OSBORN
Current results.com

Here is table 2

WHALES (5)			GLOBAL HISTORICAL CO2 EMISSION 1750 - 2020 in billion metric tons		ANNUAL AVERAGE TEMPERATURE FAHRENHEIT BY YEAR		
Year	Number of whales caught.	Rank of whales caught	Year	in billion metric tons	Year	Temperature in fahrenheit	Rank (from the highest to lowest
1880s	-	-	1889	1.25	1889	52.9	24/59
1890s	18.211	62/118	1931	4.0625	1931	49.9	52/59
1900s	-	-	1362	-	1362	-	-
1910s	-	-	1841	0.3125	1841	-	-
1920s	28.268	47/118	1927	3.75	1927	52.5	28/59
1930s	76.066	7/118	1966	12.5	1966	52.4	29.59
1940s	51,446	23/118	1938	5	1938	52.0	33/59
					1607	-	-

Sources:

5. Big Fish: A Brief History of whaling, published in National Geographic News.
6. Historical Carbon Dioxide Emission from Global Fossil Fuel Combustion and Industrial Processes from 1750 to 2020 (in billion metric tons)/Statista
   Published by IAN TISEO. May 9, 2023
7. Average Annual Temperature by year.
   National weather service.gov
   https://www.weather.gov sic Climate Book

Let's proceed with something related to deforestation by saying that, in the same way as when we were talking about what deforestation can trigger. Deforestation causes the release of CO2, while more CO2 causes the excess of CO2 in the atmosphere, leading to Earth's temperature rises, Ocean's temperature risen and that causes the sea ice to melt, reducing then the sea ice cover. The rise of ocean temperature, causes ocean acidification too, and both the melting of sea ice and acidification can affect some marine life like Krill, crabs, etc.. which are food to sea otter, even to whales. And we know how important are whales and sea otters in the fight against global warming, climate change.

For example if the sea otter's food become scarce due to acidification, or sea ice reduction, the well being of sea otter will be affected, they will become weak, thus decrease in number. And if the number of sea otters decreases, the threat to phytoplankton increases. And we know that phytoplankton play a major role in the fight against global warming, climate change.

It is the same with whale. If the amount of its food decreases, the well being of whale is threatened, thus affecting it. We know how important is the role played by whale in the fight against global warming, climate change.

In regard to what is in the pages above, it is obvious that we are under the threat of global warming, climate change. And we said in the pages above that all this started somewhere. When we read the history of whaling, it appears that the hunting of whales were done long ago. But "whaling changed irrevocably around the 17th Century, when advances in technology and increased demand for whale' products combine to elevate the practice to a major industry."[76]

In the pages above, it was said that a direct consequence of the rising of CO2, is increasingly devastating flooding.

And we know the role that whales play to fight against the rise of CO2. Based on what is in the pages above, we will look at the available

[76] Commercial Whaling 101 May 6, 2020
NRDC "http://NRDC.org/"

list of floods worldwide, starting from 14th Century to 21st. This will help us to see if there is any correlation between years before commercial whaling and the beginning to 2023.

We use here the Wikipedia's list of floods.[77] And according to it we have:

- In 14th Century, there were 2 floods:
    1. Saint Marcellus's flood, a storm tide.
    2. St. Mary Magdalene's flood.

- In 15th Century, 1 flood:
    1. The All Saints Day flood of 1436

- In 16th Century, 2 floods:
    1. 1530 St Felix's flood
    2. Mississippi River Flood on March 1543 The flooding reportedly lasted for 40 days.

- In 17th Century, 2 floods:
    1. The California Flood of 1605 that was caused by heavy rains and covered many parts of California in water.
    2. The Burchardi Flood that was a storm-tide that struck the North Sea Coast of North Frisia and Dithmarschem on the night between 11 and 12 October 1634.

In 18th Century, 8 floods:

1. Christmas Flood of 1717. Flood in Netherlands, Germany, and Scandinavia
2. Mississippi River Flood of December 1734 to June 1735. New Orleans was inundated by the flooding.
3. New Hampshire/Maine flood of October 1785
4. Great Pumpkin Flood of October 1786. Central Pennsylvania flood.

[77] List of floods - Wikipedia.

5. Mississippi River Flood of July 1788
6. Storofsen, Norway, flood of July 1789
7. Red River of the South flood of 1800

In 19th Century, 10 floods:

1. Mississippi River Flood of 1809
2. Mississippi River Flood of 1825. The flood of 1825 is the last known inundation of New Orleans due to spring flooding
3. Great Mississippi River Flood of 1844. The largest flood ever recorded on the Missouri River and upper Mississippi River in terms of discharge.
4. Great Mississippi River Flood of 1851. The flood occurred after record-setting rainfalls across the U.S Midwest and Plains from May to August 1851
5. The Great flood of 1862. Struck the West Coast of Norht America in December 1861 and January 1862.
6. The 1872 Baltic Sea flood. Storm surge that affected the Baltic Sea Coast from Danmark to Pomerania on the night of 12-13 November 1872.
7. Great Mississippi River Flood of 1874. Heavy Spring rains caused the Mississippi River to over flow.
8. Mississippi River Flood of 1882. Intense Spring rain storms, beginning on 19 February 1882, led to a rapid rise of the Ohio River and flooding along the river from Cincinnati to St Louis.
9. In 1889, the South Fork Dam broke, causing the massive Johnstown Flood of 1889.

In 20th Century, a total of 62 floods:

- 1910's, 5 floods:

1. In January 1910, large areas of Paris were flooded when the river burst its banks.

2. In June 1910 heavy rains caused extreme flooding throughout Central Europe.
3. The Great Flood of 1913, which included the Great Dayton Flood.
4. The 1916 Clermont, Queensland flood.
5. The Hatfield Flood of San Diego, United States.

- 1920's 4 floods:

1. In 1920, The Great Flood of TOKYO
2. The Great Mississippi Flood of 1927 was one of the most destructive floods in United States history.
3. The Great Vermont Flood of 1927
4. The 1929 New Zealand cyclones caused the deepest flood ever.

In 1930's 4 floods:

1. The 1931 yellow River flood, caused deaths in China.
2. The Ohio River flood 1937 occurred in late January and February 1937
3. The Los Angeles flood of 1938 occurred from late February to early March 1938
4. The 1938 Hansshin flood occurred in July 1938 in Japan

1940's 2 floods:

1. The 1948 Columbia River flood
2. The 1948 Berwickshire flood occurred on 12 August, when extremely heavy rain for the preceding six days caused the rivers Tweed...rainfall flooded the city

1950's 9 floods:

1. The Lynmouth flood of 1952 in United Kingdom.
2. The North Sea Flood of 1953 in United Kingdom
3. 1953 North Kyushu Flood, in Japan

4. On 15 October 1954, Hurricane Hazel struck Toronto, Ontario, Canada. The resulting rainfall flooded the city.
5. The Hunter Valley floods of 1955 in New South Wales (Australia)
6. 1957 Isahaya flood, due to heavy torrential massive rain, Japan.
7. 1957, the storm surge from Hurricane Audrey flooded southwest Louisiana.
8. In 1957, the Great flood of Valencia in Spain.
9. In 1959, the Rio Negro flooded the Central region of Uruguay.

- 1960 s 8 floods:
    1. In the year 1960, flooding from a deadly Tsunami caused by the Great Chilean earthquake in Chile
    2. The North sea flood of 1962, England, Germany, and Southern Denmark.
    3. The Vale's floods on September 25, 1962 in towns of Catalonia
    4. On 25 October 1964, in Zagreb, Croatia
    5. An earthquake in 1965 triggered the failure of six dams, two of which released millions of tonnes of water and debris, in Chilie.
    6. In 1965, Hurricane Betsy flooded large areas of New Orleans, Louisiana, United States, for up to 10 days.
    7. In 1966, the flood of the Arno River, in Florence, Italy.
    8. 1967 Uetsu flood, precipitation for four days, affecting an autumn rain front line, Japan.

- 1970 s 6 floods:
    1. On the night of 9 June 1972 the people of Rapid City, South Dakota in the United States were struck by a deadly flood that lasted two days.

2. On Luzon Island Flood in the Philippines, the "Great Luzon Flood "of 1972 was triggered by a series of storms during the 1972 Pacific Typhoon season, in Philippine.
3. In 1974, the dying cyclone Wanda triggered major flooding in Bribane, Australia.
4. On 2 July 1975, many areas of Romanian Cuverture Charpatinas were struck by major flooding.
5. In August 1975, the Bangiao Dam in China breaks apart under excess rainfall
6. On July 31, 1976, a nearly stationary thunderstorm caused a major flash flood, Colorado

- 1980 s 6 floods:
    1. During the 1980 s the Great Salt Lake reached record high water levels due to a large amount of rain and its lack of an outlet.
    2. The South African town of Laingsburg was basically destroyed on 25 January 1981.
    3. On August of 1982, Ulaanbaatar experienced a major flood that swept most of the houses.
    4. In 1982, the river Jucar in Spain breaks the Tous Reservoir, flooding the surrounding land in a deluge of 16,000 m3/s of water.
    5. In the Winter of 1983, the Pacific North-West of the United States, saw one of the worst floods on record for that region.
    6. Bangladesh floods.

- 1990-2000, 18 floods:
    1. January 1992 saw severe floods in South America, most notably Brazil.
    2. In Alaska, United States, from May to September 1992 it was unusually wet, causing the 100 year flood. Snow melt only made the flood.

3. The Great Flood of 1993 was one of the most destructive floods in United States history.
4. March 1993 the "No Name" storm, silently brought major flooding to Citrus county, Florida.
5. The Summer of 1993 was unusually for the United States.
6. The 1993 India floods, in Northern India, Nepal, with fewer casualities in Bangladesh and Pakistan.
7. 1994 South Georgina floods
8. On 8 May 1995, severe floods caused extensive damage in Louisiana, United States.
9. 1994- August, Massive Flood in Surat.
10. 1996 Pacific Northwest floods.
11. A dying typhoon hit Kyushu, Japan, in September 1996, causing severe floods in that region.
12. July 1996 saw severe floods in Central Honshu^, Japan.
13. In August 1996, flood in Las Nieves camping, Spain.
14. 1997 Central European flood, the worst flood in Polish history in July 1997.
15. The Red River Flood in 1997, United States in April and May 1997 (in North Dakota, Minnesota and Manitoba
16. Bangladesh was flooded in 1998
17. The 1999 Pentecost flood (German) was a 100-year flood around the Pentecost season in 1999. It was caused by heavy rains followed by a cyclone, covered much of the country for three weeks.

- 21st Century: from January 2001 to December 2100, up to 2023, 176 floods:
  1. 2000s, 0 flood.
  2. 2001, 1 flood
  3. In June 2001, floods from Tropical Allison, Houston, Texas
  4. 2002 - 2003, 4 floods
     - The 2002 northern Chile floods and Mudflow, in Chile in June 2002

- In 2002, the 2002 Glasgow floods, Scotland.
- The 2002, the 2002 European floods, Central Europe
- On October 22, 2003, a large part of Western Sahara was hit by torrential rains causing flooding

- 2004-2005, 7 floods:
    1. The 2004 Bosca-stle flood on 16 August, United Kingdom
    2. In January 2005, flooding on the rivers Eden, Kent, Derwent, Greta and Cocker, England.
    3. One of Canada's most devastating floods occurred in Southern Alberta in June 2005
    4. Flooding in Mumbai, India, in July 2005.
    5. Eighty percent of New Orleans, Louisiana, United States, was flooded due to the failure of several levees on 29 August 2005 during Hurricane Katrina.
    6. Record rain across eastern Europe in August 2005 caused very severe flooding.
    7. In November 2005, in the India States of Tamil Nadu and Andhra Pradesh, heavy rains caused by low-pressure areas in the Bay of Bengal.

- 2006-2007, 11 floods:
    1. From April 19 to 21 of 2006, a large part of Namibia's Coast was hit by floods.
    2. Korea (both North and South Korea) saw one of its worse floods ever in May 2006
    3. The Mid-Atlantic States flood of 2006 in the eastern United States is considered to be the worst in that region since the flooding caused by Hurricane David in 1979.
    4. Ethiopia saw one of its worst floods ever in August 2006.
    5. Surat, a city of 5 million people in India, witnessed the largest flood in its history during 4 to 10 August 2006. Water discharged for 30 hours from Ukai dam, which flooded the city.

6. Peninsular Malaysia, Sumatra, and Sabah suffered floods between December 2006 and January 2007. It was the worst flood in Malaysia for over 100 years.
7. The 2007 Hunter Floods, in Australia in June 2007.
8. Between late May 2007 and early August 2007, severe flash floods hit most of the United Kingdom.
9. The 2007 Africa Floods was one of the worst and most destructive floods in recorded history on the continent of Africa with 14 countries affected.
10. In November 2007, cyclone Guba, a slow moving storm caused deadly flooding in Papua New Guinea.
11. The 2008 Indian floods affected several States in India between July 2008 and September 2008 during an unusually wet monsoon season.

- 2008-2009, 4 floods:
    1. 2008 Santa Catarina floods.
    2. 2009 Brazilian floods and mudslides.
    3. In June 2009,minor flooding hit parts of Sheffield City Center in Sheffield, England. Waters reached only about half a foot deep as the River Don broke its banks.
    4. In November 2009, record-breaking amounts of rain were dumped on Cumbria, England and Cork, Ireland.

- 2010, 10 floods:
    1. January 2010 Rio de Janeiro floods and mudslides.
    2. Rio de Janeiro had its worst ever flood in April 2010.
    3. 2010 northeastern Brazil floods
    4. Between June 2010 and August 2010, flooding in China affected more than 230 million people
    5. On 26 July 2010, Heavy monsoon rains flooded most of Pakistan in 2010 Pakistan floods
    6. On 4 August 2010, at 9:25 am EST a major thunderstorm producing large hail and winds in excess of 60 mph (97 km/h)

advanced at the leading edge of a cold front moving across the American Midwest, causing a flash flood that struck Louisville, Kentucky, and portions of the surrounding Kentuckiana region.

7. In November 2010, many areas of Cornwall, UK, were struck by floods.
8. The November 2010 Colombia floods associated landslides.
9. The November 2010 Thailand floods and 2010 north Malaysian floods.
10. The 2010-2011 Queensland floods are some of the worst the country of Australia has ever seen.

- 2010 s

  - 2011, 7 floods:

1. The January 2011 Brazil floods are considered the worst in the country's history. As of 18 January 2011, the floods had taken about 700 lives... due to landslides.
2. The Mississippi River floods in April and May 2011 were among the largest and most damaging **recorded along the US waterway.**
3. **In June 2011, the Thailand floods spread through the provinces of Northern, Northeastern and Central Thailand along the Mekong and Chao Phraya river basins and persisted in some areas until mid-January 2012.**
4. **In August-September 2011, there was floods in Khammouane Province in Laos, and then came to Cambodia, and was drained via Mekong river to Vietnam then South China Sea.**
5. **On 18 October 2011, Pulau Tioman of Malaysia was flooded in Kampung Tekek, and the jungle near the village, then came up to Salang on 23 October 2011, the drain was started in November.**

6. Around November 2011, northern part of Malaysia was flooded, and then to Narathiwat Province in Thailand.

- 2012, 2 floods:
  1. In July 2012, heavy torrential rains caused floods in Kyusu, Japan.
  2. In 2012 Great Britain and Ireland floods caused many floods in the united Kingdom, in April floods and gales hit most of England causing flooding and power outages, on 28 June 2012 there were two severe supercell thunderstorm which traveled across the west Midlands causing flash flooding, on 6 July 2012 heavy rainfall brought floods to the South West of England with the Met office issuing red rain warnings, flooding later returned to the UK on 23 November 2012, as heavy persistent rainfall fell in South West England which caused rivers to burst their banks, the rain later pushed into the Midlands overnight causing more flooding, on 26 November 2012 another bank of rain pushed into South West England.

- 2013, 7 floods:
  1. The 2013 North India floods and landslides caused by heavy rainfall.
  2. The 2013 European floods.
  3. 2013 Alberta flood - on 20 June 2013, widespread flooding in Southern Alberta, when the Cougar creek, Highwood River, and other rivers and creeks overflowed caused by extensive rainfall. Flooding also caused power outages and the closure of the Trans - Canada Highway and Highway 1A.
  4. The 2013 Southwest China floods.
  5. The 2013 Afghanistan - Pakistan floods.
  6. The 2013 Colorado floods.
  7. On 18 November 2013 the heavy flood caused by cyclone Cleopatra in the Italian Island of Sardinia.

- 2014, 9 floods:
    1. Many parts of the United Kingdom experienced flooding at the start of this year. In January and then again in February the River Thames breached its banks resulting in severe flooding to many homes...
    2. In early April 2014, cyclone It caused disastrous flooding.
    3. Between 2 and 30 April 2014, flood events in the United States, caused by an important tornado outbreak.
    4. In May 2014, multiple floods affected a large area of Southeastern Europe. A low pressure area named "Yvette "brought flooding from 14 to 16 May. Bosnia, Serbia and Romania were hit by the biggest flood in their modern history.
    5. In June 2014, a flash flood in the Baghlan province of Afghanistan...
    6. 2014 Alberta floods: on 18 June 2014 the city of Claresholm, Alberta awoke to find its city streets flooded...
    7. 2014 India - Pakistan floods: In September 2014, an estimated 557 people died in India and Pakistan as a result of flooding, which was caused by intense rainfall in the area.
    8. 2014 New York flood: record setting rainfall creating 60 days of precipitation fell near New York.
    9. At the end of November 2014, a large part of Morocco was hit by floods.

- 2015, 7 floods:
    1. 2015 Northern Chile floods and Mudflow
    2. 2015 Tbilisi flood
    3. 2015 Mississippi floods
    4. 2015 Houston - Memorial Day Flood
    5. 2015 South India floods
    6. 2015 Poland Flood
    7. 2015 Myanmar Flood

- 2016, 9 floods:
    1. 2016 Sao Paulo flood and mudslide
    2. 2016 Houston Tax Day floods
    3. 2016 Ethiopia flood
    4. 2016 European floods
    5. 2016 Oklahoma floods
    6. 2016 Maryland flood
    7. 2016 Niger flood
    8. 2016 Louisiana floods
    9. 2016 Johannesburg flood

- 2017, 10 floods:
    1. 2017 Benue State flooding
    2. 2017 China floods
    3. 2017 Southern Thailand floods
    4. 2017 Peru floods
    5. 2017 Payson flash floods
    6. 2017 Quenec floods
    7. 2017 West Africa floods
    8. 2017 Hurricane Harvey
    9. 2017 Flood of Chennai in Tamil Nadu

- 2018, 8 floods:
    1. 2018 EastAfrica floods
    2. 2018 Japan floods
    3. 2018 Kerala floods
    4. 2018 Vietnam floods
    5. 2018 North Korean floods
    6. 2018 European floods
    7. 2018 Maryland flood
    8. 2018 Punjab flood

- 2019:
    1. 2019 Iran floods
    2. March 2019 North Iran floods
    3. 2019 Midwestern U.S floods
    4. South Sulawesi floods
    5. 2019 Townsville flood
    6. 2019 Pakistan floods and storms
    7. May 2019 Houston Flash Flooding
    8. June 2019 Southern and Southeastern U.S flooding
    9. 2019 Indian floods
    10. Venice floods
    11. 2019 Israel flash flood
    12. 2019 Ubon Ratchathani, Thailand flood
    13. England floods
    14. 2019 Quebec, Ontario and New Brunswick floods
    15. 2019 - 2020 Congo River floods

2020, 21 floods:

1. 2019-2020 Congo River floods
2. 2020 Jakarta floods
3. 2020 China floods
4. 2020 Kyushu floods
5. 2020 Nepal floods
6. 2020 Brazilian floods and mudslides
7. 2019-2020 United Kingdom floods
8. Storm Gloria (Spain and France)
9. Southeast Kentucky floods of 2020
10. 2020 East Africa floods
11. 2020 Assam floods
12. 2020 Korean floods
13. 2020 Afghanistan flood
14. 2020 Kerala floods
15. 2020 Karachi floods
16. 2020 Vidarbha floods

17. 2020 African Sahel floods
18. 2020 Yemen flood
19. 2020 Central Vietnam floods
20. 2020 Hyderabad floods
21. 2020 Zagreb flood

- 2021, 22 floods:
    1. 2021 Evros floods
    2. 2021 Uttarakland floods
    3. 2021 Eastern Australia floods
    4. 2021 Central New Zealand floods
    5. 2021 European floods
    6. 2021 Henan floods
    7. 2021 Oman floods
    8. 2021 Yalta floods
    9. 2021 Arizona floods
    10. 2021 Mumbai floods
    11. 2021 Lagos floods
    12. 2021 Hulunbuir dam collapse
    13. 2021 Maharashtra floods
    14. 2021 China floods
    15. 2021 Turkey floods
    16. 2021 August japan floods
    17. 2021 Nuristan floods
    18. 2021 Afghanistan flood
    19. 2021 Tennessee floods
    20. 2021 Niger floods
    21. 2021 Bosnia and Herzegovina floods
    22. November 2021 Pacific Northwest floods

- 2022, 18 floods:
    1. 2022 Petropolis floods
    2. 2022 Eastern Australia floods
    3. 2022 Assam floods

4. 2022 Bukavu floods
5. March 2022 Suriname flooding
6. 2022 Montana floods
7. 2022 Missouri floods
8. 2022 Nigeria floods
9. 2022 Kentucky floods
10. 2022 Iranian floods
11. 2022 China floods
12. 2022 Texas floods
13. 2022 Mississippi floods
14. 2022 Pakistan floods
15. January 2022 Montevideo Flood
16. 2022 Florida floods
17. 2022 South Carolina floods
18. 2022 Saudi Arabia floods

- 2023, 4 floods:

1. 2022-2023 California floods
2. 2023 North India floods
3. 2023 South Korea floods
4. 2023 Slovenia floods.

After having gone through the available list of floods in the preceding lines, we can come to the following information:

- from the 20th Century B.C to 17th Century, it must be 3,700 years, the world had recorded eight floods.
- While from the 18th Century to 2023 (21st Century), it must be 323 years, the world has recorded 256 floods, it must be 32 times the number of floods recorded in 3,700 years. We will come back to that, but let us look at the floods by region.

FLOODS BY REGION

Africa = 17
Asia= 88
Europe = 47
North America = 35
Oceania =23
South America = 12

If we bring in here the top 4 continents in CO2 emission from 1750 to present, we will have: Asia, North America, Europe, and then East Asia and the Pacific.[78]

From the information above there is a relative correlation between the high ranking CO2 emitter's continents and the high ranking continents in number of floods, though "CO2 emissions from a single, localized area can not necessarily cause flooding in that exact spot, because there are other factors that come in for flooding to happen. However, increased CO2 emissions on a global scale contribute to global warming, climate change, which can lead to more frequent and intense rainfall events, thereby increasing the risk of flooding in various regions, including the area where the emissions originated.

But the emissions of CO2 from any specific localized area affect the all world, while the specific area will be affected from it too. That is why global warming, climate change being a world threat, require that the world put efforts together to address efficiently the threat.

To proceed in the same way, remember that in the pages above it is said that a direct consequence of the rising CO2, is increasing devastating flooding. And we know that the rise of CO2 is shown by scientists, to have originated from or around the industrial revolution's era, that is around 18th Century.

Here we put, classify the list of floods quoted above, in two groups:

1. From 20th Century B.C to 17th Century, 3,700 years. And

[78] Al Overview - Google search

2. From 18th Century - the industrial revolution era - to 2023 (21st Century) 323 years. We remind here that besides the CO2 rise that leads to heavy rainfall, there are other factors like: collapsed dams, high density of people and poor management of the located area, storm surge... causing flooding. However, when we look at the list of floods above, we can observe a correlation between the period going from the industrial revolution's era to 2023, and a huge number of floods, 256 floods within 323 years. This is 32 times the number of floods recorded in within 3,700 years during the period from 20th Century B.C to 17th Century. This shows the influence of human action on flooding, which is one of the phenomena of global warming, climate change.

In the pages above we talk about flooding as a disaster triggered by the rise of CO2 in atmosphere as one of the causes. Now we will talk about cyclones, a disaster caused by the rise of CO2 too, along with other causes. Here someone may asks if Co2 emissions can cause cyclone? The answer is "yes, climate change is causing cyclones to become more intense and destructive. The oceans have absorbed 90% of the heat caused by greenhouse gas emissions, which is warming the oceans and fueling the intensity of storms. Warmer water fuels cyclones and gives them their strength. Climate change creates the conditions, in which more powerful storms can form."[79]

In the following lines, we will focus on tropical cyclones that are generally considered more dangerous, destructive than the other types of cyclones, like extra tropical cyclones (temperate cyclone or mid - latitude cyclones). Based on the table of tropical cyclones by year - Wikipedia(80) , we have divided the tropical cyclones in two groups:

1. From the year 957 to 1700, it must be 743 years. There were a total of 247 tropical cyclones. In our subdivision, this period is corresponding to the pres - industrial era.

---

[79] Al Overviews are experimental.
Is the cyclone caused by the CO2 emissions - Google Search.

2. From 1701 to 2022, 321 years. There were a total of 13,191 tropical cyclones. This period goes from or around the industrial era.

When we look at the two groups above, it appears that group one: from the year 957 to 1700, 743 years long has a total of 247 tropical cyclones. While the second group: from 1701 to 2022, 321 years long - not reaching even the half of the first group's number of years - has recorded 13,191 tropical cyclones, more than 53 times the number of cyclones recorded in the first group.

When we reflect this to the carbon dioxide emissions 1750 - 2021, and to the forest's loss 1700 to 2018 as given in the pages above.

- For carbon dioxide emissions: it is from around 1850, that the line graph started ascending, showing the CO2 emission's increase, going from around zero in 1850 to 34 billion metric tons in 2020.
- When it comes to the forest's loss in 1700, the forest was 52% of the total land, 50% in 1800, 48% in 1900, 44% in 1950 and 38% in 2018. This means that the world's forest's loss generally increased in the period going from 1700 years to 2018. With all this taken in account, it appears a correlation from one hand between the rise of CO2 emissions and the increase number of tropical cyclones. And in the other hand, the increase number of the forest's loss and the increase number of tropical cyclones.

Let us look at the data in the comparative table (8 of historic's most Catastrophic Floods, Global Land and Ocean May Temperature Anomalies (1901 - 2000 mean), ten warmest years (1880 - 2022), Global Mean annual temperature average per decade, Global historical CO2 emission 1750 - 2020, number of whales caught and Annual average temperature by year. This table is on page 32.

When looking at the data from this table, it appears the following: out of 8 historic's most catastrophic floods happened when the global land and ocean May temperature's ranking was within the 24th to 92nd out of 174 going from less to high. When looking at the global land and ocean May temperature anomaly's ranking, except two; one ranking 24th and the other one 38th, all the other 3 are ranking within 63rd to 92nd out of 174. None of the five floods is below

24th ranking in global land and ocean May temperature anomalies. Thus we can observe a correlation between the happening of the majority of the 8 historic's most catastrophic floods and the period when the global land and ocean May temperature anomalies are happening.

As we know, the rise of temperatures is mostly caused by the CO2's emissions in atmosphere. And none of the 8 historic's most catastrophic floods happened during the ten warmest years (1880 - 2022). However the line graph of the global historical CO2 emissions 1750 - 2020,[80] shows a rise of CO2 emissions from around 1850 to 2021. And looking at the 8 historic's most catastrophic floods, 75% happened within 1850 and 2021.

When we look at the number of whales caught, and that we relate it to the 8 historic most catastrophic floods, it is for four years that we have data available. Among the four years, when looking at the ranking of whales caught (from high to low), only one is ranking 62nd out of 118, and the other 3, rank within the 7th to 47th out of 118. Based on the four years with data available, there might be some correlation between the loss of whales and the rise number of historic most catastrophic floods.

When it comes to global historical CO2 emissions 1750 - 2020, as said in the pages above, from or around 1850 - 2021, the CO2 emission's line graph shows in general a rise of CO2 emissions. And out of 8 historic most catastrophic floods, 6 or 75% happened from 1850 years.

For annual average temperature by year (from the highest to the lowest), when we connect to the 8 historic's most catastrophic floods, we have five years with available data. For these five years, only one

[80] Historic Carbon Dioxide Emissions from Global Fossil Fuel Combustion and Industrial Process from 1750 to 2020. Published by Ian Tiseo, May 9, 2023

ranks 52nd out of 59, meaning among the lowest average ranking. And among the four, none ranks 1st to 23rd top ranking. However, for all the five, except two that rank 52nd and 33rd out of 59, the three are among the first half top high average ranking.

Though we can not observe a strong correlation, some measure of correlation is not to be excluded. Thus, it appears some correlation between the annual average temperature by year and the 8 historic's most catastrophic floods.

Still talking about the influence of man on the global warming, climate change. "In the cement manufacturing too, carbon dioxide is being released."[81] And cement has been used since long ago. In the preceding pages we have said that on the one hand we have overloaded the atmosphere, the planet with polluters. And in the other hand, we have destroyed significantly the natural carbon sinks, the ocean, the forest, etc... It is in this way that we talk about acres of forests being destroyed around the planet.

Now let talk about the ocean. The water's body covers almost three quarter of the all planet,and the ocean is part of it. As we are talking about ocean as a natural carbon's sink, it is relevant to know that when stable, naturally the ocean has the ability to suck in carbon dioxide and store it.

And we can read: "the ocean has absorbed about 29 percent of global CO2 emissions since the end of the pre-industrial era.[82]

"Carbon cycle experts estimate that natural sinks - process that remove carbon from the atmosphere on land and in the ocean - absorbed the equivalent of about half of the carbon dioxide we emitted each year in the 2011-2020."[83]

---

[81] Reports & Multimedia/Explainer. How do we know that Humans are the Major cause of Global warming? Published Jul 14, 2009. Updated Jan 21, 2021.

[82] Henry Shue: professor of politics and International Relations at the University of Oxford. In Got Science? PODCAST Sarah Cooley.

[83] Climate change: Atmospheric Carbon Dioxide.
By REBECCA LINDSEY/Reviewed by ED DLUGOKENCKY
Published May 12, 2023

However, too much carbon dioxide, beyond the ocean absorption's capacity, becomes a harm to the natural balance of the planet. And in the pages above, we have said that since the industrial revolution, we have dropped tons and tons of carbon dioxide into the atmosphere. That is how the human over emissions of carbon dioxide, affect negatively the ocean absorption ability, slowing down its carbon dioxide's absorption.

This is because "the rising concentration of carbon dioxide in the atmosphere is driving up ocean surface temperatures and causing ocean acidification. Although warming and acidification are different phenomena, they interact to the detriment of marine ecosystems."[84]

And we remember as it was already said above that "since the beginning of the industrial revolution, we have dumped tons of carbon dioxide in the atmosphere. And by then the PH of the ocean's surface waters has dropped from 8.21 to 8.10. This drop in PH is called ocean acidification.[85]

While the ocean is a natural carbon dioxide sink on its self, there are two organisms living in the sea, which play as well a major role in the fight against global warming, climate change. We are talking about whale - a great mammal - and phytoplankton, a microscopic organism.

When talking about whales: "live or dead of natural causes, they play a major role in the removal of tons of carbon dioxide. While dead, whales carry the tons of carbon stored in their massive bodies down to rest on the sea floor where it remains for centuries. Alive, whales fertilize the ocean with their feces and urine which make nutrient and leading large phytoplankton blooms. Phytoplankton produce at least 50%

[84] Ibidem; Henry Shue: professor of politics and International Relations at the University of Oxford. In Got Science? PODCAST. Sarah Cooley

[85] Ibidem: Climate Change: Atmospheric Carbon Dioxide by REBECCA LINDSEY/Reviewed by ED DLUGOKENCKY Published May 12, 2023.

of the oxygen in the atmosphere and capture an estimate 40% of all carbon dioxide produced in the world, according to the IMF report."[86]

But when killed to be eaten or used for other purposes, the carbon dioxide that was stored in whales, return into the atmosphere just like when trees that are cut down release the carbon dioxide they kept in them, into the atmosphere. According to the 2019 report in the magazine of the International Monetary Fund (IMF), "each great whale sequesters 33 tons of carbon dioxide on average taking that carbon out of the atmosphere for centuries."[87]

And if we take the statistics about the number of whales during the preindustrial era, the statistics say there were "4 to 5 million of whales."[88] With these statistics if whales can recover to reach an average of 3 millions of whales able to sequester carbon dioxide, 99.000,000 tons of carbon dioxide would be removed from the atmosphere.

Besides that, we can not imagine the amount of phytoplankton which would be increased, boosted by the action of the whale's feces, urine and its up and down movements that enhance the nutrients, which are foods to phytoplankton. Phytoplankton - as said above - produce at least 50% of the oxygen in the atmosphere and capture an estimate of 40% of all the carbon dioxide produced in the world.

Unfortunately, whales have been killed in large number to near extinction since the industrial revolution, depriving to our planet a key solution to the world's threat, the global warming, climate change.

"Biologists estimate that there are now fewer than a quarter of the whales that once existed, according to whales experts Mongabay

---

[86] MONGABAY, News & Inspirations from nature's frontline. To fight Climate Change, save the whales, some scientists say. By Michelle Carrere
On 1 March 2021/Translated by Sarah Engel.

[87] https://www.imf.org/external/pubs/ft/food /2019/12/natures - solution - to - climate - chami. Htm

[88] Ibidem: MONGABAY, News & Inspirations from nature's frontline. To fight Climate Change, Save the whales, some scientists say. By Michelle Carrere
On 1 March 2021/Translated by Sarah Engel.

interviewed for this story."[89] And that is a bad news concerning the fight against global warming, climate change, because: "Whales both increase the ocean ability to absorb carbon dioxide and at the same time actually act as living sinks that store carbon dioxide in their bodies.[90]

When it comes to phytoplankton, as already said above, this marvelous microscopic organism suck in carbon dioxide from the atmosphere through photosynthesis process to produce oxygen. Compare to the trees, forests, "phytoplankton captures approximately four Amazon rainforests worth of carbon dioxide each year."[91] So the increase of phytoplankton is a great good news because it is beneficial as far as the fight against global warming, climate change are concerned.

But while we wish phytoplankton to flourish in the amount needed to fight global warming, climate change, Urchins act against the phytoplankton's growth, because phytoplankton is food to urchins. Urchins destroy phytoplankton on a daily basis.

Fortunately, while sea urchins eat phytoplankton, sea otter - a marine mammal - eat sea urchins and this leads to phytoplankton to flourish and by then play its role of sucking in, tons of carbon dioxide. While whale and phytoplankton have a direct influence in the fight against global warming, climate change, sea otter on the other hand, plays an indirect role. So the importance of sea otter in this fight should not be neglected.

But like whales, sea otters have been subject to massive killing since the industrial era. For example, "from 1990 s to 2005, the population of the northern sea otters plummeted by 90 percent off the

[89] Protecting Whales to protect the planet. Friday Jul 21, 2025 at 5:07 pm

[90] Why whales Are Important for Carbon sequestration.
Francisco Basetti August 27,2020

[91] Idem: Why whales are important for carbon sequestration.
Francisco Basetti. August 27, 2020

Central West Aleutian Islands that was once home to around 80 percent of all the sea otters around the World."[92]

And the decrease of sea otter population means the sprouting of sea urchin which is a threat to the sprout of phytoplankton. And the importance of phytoplankton in the fight against global warming, climate change is not negligible. That is how in one hand, the human activities harm to the natural carbon dioxide's capacity, and in the other hand, years after years human keeps accumulating the amount of carbon dioxide by dumping million tons of it in the atmosphere. This accumulations lead to the excess amounts of carbon dioxide in atmosphere causing global warming, climate change as this excess 'greatly amplify the initial solar - driven warming."[93]

In the pages above, based on the structural-functional approach, we said that the planet is like a system composed of different elements in interaction. The malfunction of an element may affect other elements and lead to the malfunction of the all system.

We can remember that the excess of carbon dioxide in the atmosphere is one of the main causes of temperature's rises and thus the cause of global warming, climate change. And these rises of temperature lead to other phenomena that affect the natural balance, the well being of life in the planet. In this regard, the continue rise of temperature may cause or lead to permafrost melting. "Permafrost is a combination of soil, sediment and dead plants and animals remains that stay at or below 0 degree Celsius for at least two years."[94]

---

[92] How many Sea otters Are Left in the World? Sea Otter Population. Animals answers
By UBAID KHALID November 8, 2017

[93] Ibidem; Climate Change: Atmospheric Carbon Dioxide
By REBECCA LINDSEY/Reviewed by ED DLUGOKENCKY
Published May 12, 2023

[94] Global Surface Ocean Acidification Indicators. From 1750 to 2100 - Jiang - 2023 - Journal of Advances in Modeling Earth Systems
AGU Advancing Earth And Space Sciences
Volume 15, ISSUE 3 March 2023 e2022MS003563
JAMES JOURNAL of Advances in Modeling Earth System

Permafrost "covers 24 percent of the surface of land masses in the northern hemisphere and accounts for nearly half of all organic carbon stored within the planet's soil."[95] Frozen ground permafrost is very sensitive to the rises of temperatures, that can lead to its thawing. "With thawing, microbes will begin to eat the material, causing it to decay and releasing carbon dioxide and methane into the atmosphere. Even if a small fraction of these greenhouse gases are released, it will have consequences on not only the Arctic, but Earth's entire climate system, as they intensify global climate change. And while the Arctic was previously considered a carbon sink, new research, shows that the region is emitting more carbon than it is absorbing, largely due to permafrost thaw. It is estimated that the world permafrost contains up to 1,700 billion tonnes of carbon, which is almost double the amount of carbon in the Earth atmosphere."[96]

This is for the Arctic region only. Though most of the permafrost in the planet lies in the Arctic, the Antarctic has some too.

Recent observations "show parts of the Arctic warming 2-3 times faster than the rest of the planet."[97]

What is in the lines above, means that when we are at a 1.5 degree Celsius above the pre-industrial average, in those parts of Arctic, it is around 3 to 4.5 degree Celsius more than the temperature in the pre-industrial era. And we can see how big is the threat, knowing that with what we have just said in the lines above, thawing is a threat and its negative impacts.

---

[95] Permafrost Thawing in a Warming World: The Arctic Institute's Permafrost Series
Fall - Winter 2020/The Arctic Institute's Permasfrost Series

[96] Idem: Permafrost Thawing in a warming World: The Arctic Institute's Permafrost series

[97] Global Surface Ocean Acidification Indicators. From 1750 to 2100 - Jiang - 2023 - Journal of Advances in Modeling Earth System
AGU ADVANCING EARTH AND SPACE SCIENCES
Volume 15, ISSUE 3 March 2023 e2022MS003563
JAMES JOURNAL of Advances in Modeling Earth System

The threat is real when we consider the following: "buried in the circumpolar region of the Arctic are 1.4 trillions of carbon, two times more than in all the forests on the planet,"[98]

Besides that, thawing permafrost put more greenhouse gases (carbon dioxide, methane, nitrous) into the atmosphere, it causes the ground to loose its stability, damaging the infrastructure, the ecosystem, thus affecting communities living in the region. And "more than half of all the land in the Northern Hemisphere freezes and thaws every year."[99]

All this started somewhere: human activities put more greenhouse gases in the atmosphere and that led to the rises of temperatures. The rises of temperatures interact to cause the permafrost thawing, and the permafrost thawing causes the release of more greenhouse gases. And more greenhouse gases fuel global warming, climate change, and the all system is affected. There are clathrates too, which are found both in Arctic and Antarctic. From google, "clathrate is defined as a substance with holes of appropriate size inside the three - dimensional structure formed by bonding of atoms or molecules, in which other atoms or molecules exist at a fixed composition ratio. And scientists believe they could hold anywhere from 100,000 to 1 million Tcf (Trillion Cubic feet)"[100]

On Earth, gas clathrates can be found beneath the sea floor where there is enough pressure to keep their molecular structure stable. "The United States Geological Service estimate the amount of carbon in

[98] DRAWDOWN. THE MOST COMPREHENSIVE PLAN EVER PROPOSED TO REVERSE GLOBAL WARMING
Edited by PAUL HAWKEN
PENGUIN BOOKS
An imprint of Penguin Random House LLC
375 Hudson Street New York, New York 10014 Penguin.com
Copyright © 2017 by Project Drawdown
"Reciprocity" by Janine Benyus © 2017 Janine Benyus. Published by arrangement with the author.

[99] NSIDC Frozen Ground & Permafrost/National Snow and Ice Data Center

[100] Google

clathrates is twice the amount of carbon that exist in all the fossil fuels on Earth. And when clathrates melt in the ocean, some is oxidized and acidifies the ocean, while some escapes into the atmosphere."[101]

This leads us to talk about acidification, that we already talk about in the pages above. But we talk about it here as another phenomenon resulting from the dysfunction of one or some elements of the system Earth.

As already said, human activities dump more greenhouse gases, Carbon dioxide, methane, nitrous oxide into the atmosphere and that leads to the rises of temperatures. "Half of the carbon dioxide that is recaptured from the atmosphere goes into oceans, causing surface acidification. And over 90% of the heat caused by global warming is absorbed into the surface waters, a trend that is steadily erasing the marine food chain."[102]

What happens is that with this, more carbon dioxide in the atmosphere, the sea being a natural sink, carbon dioxide is absorbed, and this is good to lower the amount of carbon dioxide in the atmosphere. But the problem arise when the sea absorbs more carbon dioxide leading to acidification.

It has started somewhere: human activities dumping more greenhouse gases (carbon dioxide, methane, nitrous oxide, etc...) The accu-

---

[101] Ibidem: Global Surface ocean Acidification Indicators. From 1750 to 2100 - Jiang -2023 - Journal of Advances Modeling Earth System
AGU ADVANCING EARTH AND SCIENCES
Volume 15, ISSUE 3 March 2023 e2022MS003563
JAMES JOURNAL of ADVANCES in MODELING EARTH SYSTEMS

[102] Ibidem: DRAWDOWN. THE MOST COPREHENSIVE PLAN EVER PROPOSED TO REVERSE GLOBAL WARMING
Edited by PAUL HAWKEN
PENGUIN BOOKS
AN imprint of Penguin Random House LLC
375 Hudson Street New York, New York 10014
Penguin.com
Copyright © 2017 by Project Drawdown
"Reciprocity" by Janine Benyus © 2017 Janine Benyus.
Published by arrangement with the author.

mulation of those gases, mainly CO2 leads to the rise of the land's temperature as well as sea water's temperature, at the same time the sea absorbing more CO2. More concentration of Co2 in the sea leads to the sea acidification. Then the sea acidification, the warmer sea temperature disturb seriously the marine life, and the perturbation of the marine life, "pose a serious threat to the survival of the world's great whales."[103]

And we know the importance of whales in the fight against global warming, climate change.The same way human activities dump more greenhouse gases in the atmosphere, more greenhouse gases (carbon dioxide, methane, nitrous oxide, etc...) leads to the rises of temperatures, the over absorption of carbon dioxide in the sea leads to ocean acidification, and ocean acidification affects the sea otter's life: "shellfish have calcium - carbonate - based shells, which are quite sensitive to that level, and any increase makes it harder to grow and maintain healthy shells.

Fall - Winter 2020/The Arctic Institute's Permafrost series

Any impact on the health and number of shellfish, - as sea otter's primary prey-, translate into a potential impact on sea otters already constrained by a limited food supply."[104]

And we know that sea otters play a very important indirect role in the fight against global warming, climate change.

The same way, human activities dump tons of greenhouse gases into the atmosphere, the excess of those greenhouse gases leads to the temperature rise. More carbon dioxide absorbed by the ocean, and the rise of temperature in the ocean's water lead to ocean acidification. And the increased ocean acidification affects the life and the growth of phytoplankton: "the krill and plankton that form the foundation

---

[103] The surprising Role of whales in ocean carbon capture. Friday, July 21, 2023 at 5:11 PM

[104] The Edge Extinction: can sea otters survive human threat? - Kim Steinhardt. Kim Steinhardt (https:// coastal care.org/tag/Kim - Steinhardt/)

of the marine ecosystem are also being affected by the increasing ocean acidification. More CO2 being absorbed by the ocean means that some of these species will be unable to form and maintain their protective calcium carbonate shells, threatening key species further up the food chain."[105]

In the same way, human activities dump tons of greenhouse gases, mainly the CO2, the excess of these gases, the excess of CO2 in the atmosphere leads to the rise of temperature. The absorption of more carbon dioxide in the ocean affect the capacity of the ocean to suck carbon dioxide and this reduces the possibility of the ocean to be a natural carbon's sink, thus reducing the ocean's capacity to fight against global warming, climate change.

All this shows that the Earth as a system can be affected when one of its parts is in dysfunction. For example human being living in biosphere - a subsystem of the Earth -, the atmosphere, another subsystem become overwhelmed by the excess of the greenhouse gases,dumped by human being who is part of the biosphere. And because of those greenhouse gases in atmosphere, this sphere the Earth becomes in dysfunction. The dysfunction of the atmosphere affects the hydrosphere another subsystem of the system Earth, etc..

All this ends up fueling global warming, climate change, affecting the geosphere - as the rocks and hard parts of the Earth is affected by global warming, climate change (the rocks are subject to erosion, to exfoliation, the rocks disintegrate; the Cosmosphere - the universe surrounding the Earth is affected too. The all system is affected too, because the Earth as a whole is influenced by the interaction of its subsystems.

To end this chapter, as said in the pages above, before we come to the situation of global warming, climate change in which we are living today, it started somewhere. In one hand human had significantly polluted the atmosphere with mostly the greenhouse gases. And in the other hand we destroyed excessively the CO2 natural sinks: forests,

[105] Ibidem: The surprising Role of whales in ocean carbon capture.
Friday, July 21, 2023 at 5:11 PM

whales sea, phytoplankton that were absorbing the CO2 gas, and sea otters that play influence indirectly this CO2 absorption.

In this chapter we took time, talking about deforestation, sea ice melting, ocean acidification, permafrost thawing, etc.

Since the commercial Whaling and sea otter's hunting led to a very significant reduction or almost extinction of these precious mammals, nevertheless vital in the fight against global warming, climate change, it is important to talk about the whale and sea otter's threat. That is why in the following chapter, Chapter 3, we will talk about human action and the threat or almost extinction of whales and sea otters.

Chapter 3

Human action and the extinction of Whales and Sea Otters

After chapter 2 which talks about human's influence as a cause of global warming/climate change, comes chapter 3, which talks about human's action and the extinction of whales and Sea otters. When reading in the preceding pages, we can realize that whales and Sea otters are one of the key solutions to fight against the World's threat, the global warming/climate change. Whales both increase, enhance the ocean's ability to absorb carbon dioxide and at the same time act as the living sinks that store carbon dioxide in their bodies for years. When it comes to sea otters, they play a major role by protecting phytoplankton. Sea otters eat sea urchins that destroy phytoplankton.

While whales and sea otters are vital in the fight against global warming/climate change, the two precious gifts to our planet have been subject to massive killing since industrial era. When we read the available data from experts or organization on whales, it is reported that there are now less than a quarter of whales that existed. Though that is not the case of sea otters, expert's reports show that sea otters are classified too as endangered species.

To give you idea of how horrible whales and sea otters have been killed and still threatened, we thought to have this chapter. As already said in the pages above, we have pumped excessively carbon dioxide into the atmosphere, and as if that was not enough, we have and continue to deplete the whale and sea otter's population, nevertheless vital in the fight against global warming/climate change, by absorbing the excess carbon dioxide in the atmosphere, and potentially reach the equilibrium, necessary for life, for the well being of the planet.

The hunting of whales and sea otters did not start today. The history of whaling and sea otter's hunting shows that several thousands years ago, whale and sea otter's hunting was taking place using rudimentary tools. Thus due to technology used to hunt these mammals, the number of whales or sea otters killed per year, was far below the numbers recorded starting the industrial era.

The rise's demand in whale's meat, oil and other products derived from whales and sea otters, drove the humanity to almost the brink of extinction of whales and sea otters.

In this regard, we can read: "population of whales in the world's oceans have been decimated in the last two centuries, falling from 5 million to 1.5 million today. More than 30 years after commercial whaling was largely banned in a bid to protect them, whales continue to suffer from human impacts in the form of plastic and noise pollution, injuries from ships, entanglements in fishing gear and changes to the ocean due to global warming."[106]

When it comes to sea otters, "by the end of the 19th Century, traders had killed hundreds of thousands of sea otters for their pelts, leaving perhaps few than 2,000 remaining. With so few surviving into the early 1900s there were not enough sea otters left to make commercial trade worthwhile. In 1911, with the entire species nearing extinction, sea otters finally received legal protection under the North pacific Fur Seal Convention agreement between the U.S, Russia, Japan and Great Britain. In a curious twist of fate, a small colony of Southern sea otters survived near Big Sur. A handful of locals knew of their existence but carefully protected that secret until the late 1930's. This secrecy allowed the sea otters to begin their recovery unimpeded

[106] Ibidem: The Surprising Role of Whales in Ocean carbon capture. Friday, Jul 21, 2023 at 5:11 PM

by further direct human interaction. They became public again only when the stretch of Highway 1 along the Big Sur Coast first opened to traffic in 1938. Yet the entire current population of roughly 3,000 Southern sea otters owes its existence to that tiny group."[107]

Besides hunting them, the threats to sea otters are legion: diseases, oil spills, lost nets, engines left in the sea, human competition with them for foods - shellfish, marine invertebrates etc.. - that are favorite foods to sea otters - are eaten by human too. This competition makes the sea otter's foods becoming scarce, while sea otters need a high amount of caloric intake to survive against the cold environment, to enable them to strive in such a cold environment. Pollution from land runs off into the oceans as well. That is a threat affecting their habitats, exposing sea otters to diverse infections.

The global warming/climate change's impacts affect their life's usual conditions though they might develop resilience mechanisms to adapt. "But due to the efforts of dedicated environmentalists and policy makers, sea otters rebounded in nearly two - thirds of their historical range. However sea otters are classified as endangered, still vulnerable to many anthropogenic threats, including oil spills, poaching, commercial fishing, and climate change."[108]

It is estimated that the worldwide population of sea otters "once numbered between several hundred thousand to over one million before being nearly hunted to extinction by fur traders in the 1700s and 1800s. Worldwide, numbers have slowly recovered but still stand far below original population numbers, though its death from human

[107] Extinction Exposed. The sea otter. Columbia Climate School.
Climate Earth, and Society. State of the Planet. General Earth Institute.
Extinction Exposes - The Sea Otter.
By BRIAN KATEMAN/September 12, 2011

[108] Extinction Exposed. The sea otter. Columbia Climate School. Climate Earth, and society. State of the Planet. General Earth Institute. Extinction Exposes - The sea otter.
By BRIAN KATEMAN/September 12, 2011.

conflict is slowly decreasing as a result of their, protection under the Endangered Species Act and increased regulation of fishing nets."[109]

But all account put together, the sea otter's growth has not yet recovered. "From 1990 s to 2005, the population of the Northern sea otters plummeted by 90 percent off the Central West Aleutian Islands. Aleutian Islands was once home to around 80 percent of all the sea otters around the world. Even though they still listed as endangered species thankfully, the numbers of these adorable creatures have now rebounded to reach 125,000."[110]

The sea otter Foundation & Trust, as well says: ["the Southwest Alaskan sea otter population declined by nearly 90% between 1990 and 2015. Meanwhile other populations remain stable or are growing, such as those along the Russian coast and part of California", - this is in line with what the "time line of sea otter population changes shows:

- 18th - 19th Century: intensive hunting for fur trade leads to near extinction.
- Early 20th Century: International and national protections lead to slow but steady recovery.
- Late 20 the Century: population recover in several areas, but new challenges like pollution and climate change emerge.
- 21st Century: despite stabilization, in some regions other populations face significant declines underlining the ongoing and evolving threat to sea otters."][111]

[109] Human Influence - Northern Sea otter.

[110] How many sea otters Are Left in the World?
Sea otter Population. Animals answers.
By UBAID KHALID
November 8, 2017

[111] Sea otter Foundation & Trust
https:// sea otterfoundationtrust.org
Understanding sea otter Population
Decline: causes and solutions
Jan 21, 2024

Coming back to whale's threat, in the following pages, different tables and a line graph will give you more clue on whale's threat starting from or around the industrial era.

Here below are the tables

TABLE ONE

Is on the following page.

Year	NUMBER OF FIN WHALES PRE-VIOUSLY SEEN	HUMP-BACK WHALES IN GLACIAL BAY & ICY	NUMBER OF SOU THERN RESI DENT KILLER WHALES	NUMBER OF EASTERN NORTH PACIFIC GRAY WHALES	NUMBER OF BELUGA WHALES IN COCK INLET	NORTH ATLAN TIC RIGHT WHALES POPU LATION	CUR RENT GLOBAL POPU LATION SIZE SPERM WHALES
1900	-	-					+- 1,687.5
1901-1959	No data available	No data available	No data available	No data available	No data available	No data available	No data available
1960	-	-	+- 78	-	-	-	-
1961	-	-	+- 79	-	-	-	-
1962	-	-	+- 82	-	-	-	-
1963	-	-	85	-	-	-	-
1964	-	-	+- 91	-	-	-	-
1965	-	-	+- 94	-	-	-	-
1966	-	-	+- 95	-	-	-	-

1967	-	-	+- 97	-	-	-	-
1968	-	-	+- 89	+- 13,125	-	-	-
1969	-	-	+- 81	+- 14,687.5	-	-	-
1970	-	-	80	+- 14,687.5	-	-	-
1971	-	-	+- 67	+-12,500	-	-	-
1972	-	-	+- 69	+-10,937.5	-	-	-
1973	-	-	+- 72	+-16,250	-	-	-
1974	-	-	70	+-17,500	-	-	-
1975	-	-	+- 71	+-15,000	-	-	-
1976	-	-	70	+-17,812.5	-	-	-
1977	-	-	+- 79	+-18,750	-	-	-
1978	-	-	+- 79	+-19,687.5	-	-	-
1979	-	-	+- 81	+-15,312.5	-	-	-

TABLE ONE (continued)

1980	-	-	+- 82	+-20,000	-	-	-
1981	-	-	+- 81	+-20,625	-	-	-
1982	0.89	-	+- 77	+-21,875	-	-	-
1983	1.79	-	+- 76	+-22,500	-	-	-
1984	3.57	-	+- 74	+-23,125	-	-	-
1985	5.36	40	+- 78	+-23,750	-	-	-
1986	12.5	45	+- 81	+-23,125	-	-	-

1987	21.43	61	+- 84	+-25,000	-	-	-
1988	16.66	55	85	+-26,875	-	-	-
1989	28.57	40	+- 84	+-25,000	-	-	-
1990	6.35	50	+- 89	+-21,875	-	+-265.63	-
1991	21.43	+- 53	+- 93	+-20,000	-	+-268.75	-
1992	71.43	+- 63	+- 91	+-17,187.5	-	275	-
1993	60.71	45	+- 98	+-15,625	-	+-278.12	-
1994	12.5	60	+-96	+-20,321.5	37.5/871	+-284.38	-

TABLE ONE (continued)

YEAR	NUMBER OF FIN WHALES PRE-VIOUSLY SEEN	HUMP-BACK WHALES IN GLACIAL BAY & ICY	NUMBER OF SOUTHERN RESIDENT KILLER WHALES	NUMBER OF EASTERN NORTH PACIFIC GRAY WHALES	NUMBER OF BELUGA WHALES IN COCK INLET.	NORTH ATLANTIC RIGHT WHALES POPU-LATION
1995	8.3	+- 54	+-98	20,625	12.5/737.5	+- 290.63
1996	5.37	+- 78	+- 97	+-21,250	28.125/700	
1997	12.5	80	+- 93	+-21,562.5	100/450	+- 309.38
1998	12.5	+- 93	+- 89	+-21,250	75 /437.5	+- 309.38
1999	16.7	+- 105	+- 84	+- 19,062.5	0.6849710983	+- 307.82
2000	37.25	90	+-82	+- 17,500	+-62.5/ 462.5	+- 310.94
2001	43.5	+- 99	+- 81	+- 16,250	1.56/418.75	+-331.25
2002	37.5	+-88	+- 83	+-16,562.50	0 /412.5	350
2003	37.25	+-118	+- 84	+-17,187.5	+-50/375	+- 359.38
2004	103.57	+- 139	+- 88	+- 18,125	+- 137.5 /437.5	+- 371.88
2005	105.35	145	+- 88	+- 19,375	+-231.25/ 321.875	+- 392.2
2006	175	150	90	+- 20,312.5	+- 12/325	+- 403.13
2007	162.25	160	+- 87	20,000	+-225/387.5	+- 415.63
2008	116.08	+- 159	85	+- 17,500	+- 187.5/475	437.5
2009	106.25	+- 185	+- 87	20,000	+- 266.7/412.5	+- 459.38
2010	75	+- 195	+- 86	21,250	+- 125/350	+- 478.13
2011	103.57	220	+- 89	+- 20,937.5	75/337.5	+- 476.56

2012	91.07	210	+- 83	+-24,375	+- 112.5/400	475
2013	78.57	240	+- 81	+-25,625	-	+- 471.89
2014	50	180	+- 78	+- 28,437	+- 75/356.25	462.5
2015	112.5	+- 162	+- 81	+-27.812.5	-	+-453.13
2016	85.72	+- 162	+- 83	+-25,625	125/337.5	+- 434.38
2017	-	+- 128	+- 77	25,000	-	411
2018	-	-	75	+- 22,500	37.5/212.5	-
2019	-	-	+- 73	+- 22,187.5	-	-
2020	-	-	+- 72	20,000	-	-
2021	-	-	+- 74	+- 17,187.5	100/156.25	-
2022	-	-	-	+-15,937.5	-	-

TABLE ONE (Continued)

YEAR	NUMBER OF BOWHEAD WHALES (1)	ICE COVER Km2 (1)	NUMBER OF BLUE WHALE POPU-LATION	SEI WHALE POPU-LATION SIZE	NARWHAL WHALE POPU-LATION SIZE (FEMALE) (07)	NARWHAL WHALE POPU-LATION SIZE (MALE) (07)	ESTIMATE NUMBER OF NORTH ATLANTIC COMMON MINKE WHALE
1980	-	-	100	-	-	-	-
1981	+- 0.0025	85,000	150	-	-	-	-
1982	+- 0.00125	+- 97,500	+- 170	-	-	-	-
1983	+- 0.001875	+- 102,500	+- 187.5	-	-	-	-
1984	+- 0.002188	+- 90,000	+- 195	-	-	-	-
1985	+-0.0025	+- 82,500	+- 212.5	-	-	-	-
1986	+-0.00281.3	+- 72,500	+- 237.5	-	-	-	-
1987	+- 0.0025	+- 77,500	+- 245	-	-	-	26,087
1988	+-0.00281.3	+- 70,000	+- 262.5	-	-	-	16,429
1989	+- 0.00375	+- 90,000	+- 273	-	-	-	64,247
1990	+- 0.00375	+- 86,250	+- 287.5	+- 23,750	+- 115.63	+- 115	-
1991	+- 0.00625	90,000	+- 290	+- 24,687	+- 118.75	+- 149	-
1992	+- 0.00125	+- 87,500	300	25,000	+- 120.83	+- 154.17	-

1993	0,00	+- 97,500	313	25,000	+- 118.50	+- 155	8,371
1994	0,00	+- 90,000	+- 320	+- 25,078.13	+- 122.5	+- 160	-
1995	25	+- 87,500	+- 313	+- 26,000	125	162.5	124,168
1996	6	+- 81,250	325	+- 26,000	+- 134.38	+- 165	23,522
1997	75	+- 82,500	+- 337.5	+- 27,500	+- 137.500	+- 172.5	26,718
1998	+- 0,0088	+- 83,750	325	+- 27,518.13	+- 131.25	+- 173.5	13,673
1999	+- 0,0088	+- 81,250	325	+- 29.998.33	+- 130	175	17,406
2000	+- 0,0075	82,500	337.5	+- 29.998.34	+- 127	+- 177.08	25,885
2001	15	90,000	350	+- 29.998.40	+- 138	+- 189	56,328
2002	0,01	82,500	375	30,000	+- 143.75	+- 197.5	-
2003	15	57,500	+- 389	31,000	+- 147.5	+- 209.38	19,377
2004	+- 0,025	60,000	+- 395	+- 32,000	+- 149.5	212.5	6,246
2005	+- 0,0212.5	20,000	+- 408.3	+- 32,000	+- 158.33	225	37,531
2006	+- 0,0262.5	30,000	425	+- 32,500	+- 158.33	+- 235.42	27,152
2007	+- 0,0262.5	47,500	+- 437.5	+- 32,600	+- 159.38	+- 247.5	81,664
2008	+- 0,0262.5	57,500	+- 455	+-33,750	+- 165	+- 256.25	27,390
2009	+- 0,0262.5	90,000	475	+- 33,900	+-184.38	+- 270.83	16,479
2010	+- 0,2688	47,500	+- 487	+- 34,500	+- 195.83	+- 277	10,991
2011	+- 0,2688	+-45,000	+- 480	35,000	+- 198	+- 277.5	21,218
2012	+- 0,2688	+- 82,500	-	+- 35,750	+- 190.63	+- 274	-
2013	+- 0,02688	+- 85,000	-	+- 35,800	+- 198	275	34,125
2014	-	-	-	+- 37,450	+- 190.63	+- 277.5	23,059
2015	-	-	-	+- 37,600	+- 190	+- 272.5	44,265
2016	-	-	-	+- 39,998.3	+- 183.33	+- 262.5	37,020
2017	-	-	-	+- 39,998.3	175	250	55,165
2018	-	-	-	+- 39,998.3	+- 160	225	17,792
2019	-	-	-	+- 40,000	+- 156.25	+- 218.75	15,693
2020	-	-	-	+- 40,850	+- 147.5	+- 207	-
2021	-	-	-	-	150	212.5	-
2022	-	-	-	-	+- 147.5	+- 208.33	-

TABLE TWO

YEAR	POPULATION	REGION
1982-1996	43,027	Western North Pacific
1985-1991	17,500	Antarctic
1986-1990	37,777	Eastern Tropical Pacific
1997	32,068	North-East Pacific
2002-2007	13,557	Northeast Atlantic
2003-2013	1842	Mediterranean Sea
2005-2007	5445	Eastern North Atlantic
2009-2010	224	Canary Islands
2007	13,443	North Atlantic and North Mariana Islands
2014	2281	California Current
2015-2016	380	Off Shore Ireland
2015	575	Gulf of Alaska
2016	4349	North Western Atlantic
2017	5095	Hawaiian Islands
2017-2018	1967	Gulf of Mexico

TABLE THREE

YEAR ESTIMATES	ABUNDANCE
2010	78
2011	104
2012	165
2013	158
2014	103

2015	180
2016	88
2017	122
2018	148
2019	218
2020	71

TABLE 4: NUMBER OF POPULATION WHALE RANKING FROM THE LEAST TO THE GREATEST.

NUMBER OF FIN WHALE RANKING			NUMBER OF HUMPBACK WHALE RANKING			NUMBER OF SOUTHERN RESIDENT KILLER WHALE RANKING		
YEAR	NUMBER OF WHALE	RANK	YEAR	NUMBER OF WHALE	RANK	YEAR	NUMBER OF WHALE	RANK
1982	0.89	1	1985	40	1	1971	+- 67	1
1983	1.79	2	1989	40	1	1972	+- 69	2
1984	3.57	3	1993	45	2	1974	70	3
1985	5.36	4	1986	45	2	1976	70	3
1996	5.37	5	1990	50	3	1975	+- 71	4
1990	6.35	6	1971	+- 53	4	1973	+- 72	5
1995	8.30	7	1995	+- 54	5	2020	+- 72	5
1986	12.5	8	1988	55	6	2019	+- 73	6
1998	12.5	8	1994	60	7	1984	+- 74	7
1997	12.5	8	1987	61	8	2021	+- 74	7
1994	12.5	8	1992	+- 63	9	2018	75	8
1988	16.66	9	1996	+- 78	10	1983	+- 76	9
1999	16.70	10	1997	80	11	1982	+- 77	10
1987	21.43	11	2002	+- 88	12	2017	+- 77	10

1991	21.43	11	2000	90	13	1960	+- 78	11
1989	28.57	12	1998	+- 93	14	1925	+- 78	11
2000	37.25	13	2001	+- 99	15	2014	+- 78	11
2003	37.25	13	1999	+- 105	16	1961	+- 79	12
2002	37.5	14	2003	+- 118	17	1977	+- 79	12
2001	43.5	15	2017	+- 128	18	1978	+- 79	12
2014	50	16	2004	+- 139	19	1970	80	13
1993	60.71	17	2005	145	20	1986	+- 81	14
1992	71.43	18	2006	150	21	1981	+- 81	14
2010	75	19	2008	+- 159	22	1979	+- 81	14
2013	78.57	20	2007	160	23	2001	+- 81	14
2016	85.72	21	2015	+- 162	24	2013	+- 81	14
2012	91.07	22	2016	+- 162	24	2015	+- 81	14
2011	103.57	23	2014	180	25	1969	+- 81	14
2004	103.57	23	2009	+- 185	26	1980	+- 82	15
2005	105.35	24	2010	+- 195	27	1962	+- 82	15
2009	106.25	25	2012	210	28	2000	+- 82	15
2015	112.5	26	2011	220	29	2012	+- 83	16
2008	116.08	27	2013	240	30	2002	+- 83	16
2007	162.25	28	-	-	-	1989	+- 84	17
2006	175	29	-	-	-	1987	+- 84	17
						2003	+- 84	17
						1999	+- 84	17
						1988	85	18
						1963	85	18
						2008	85	18

						2016	+- 83	16
						2010	+- 86	19
						2007	+- 87	20
						2009	+- 87	20
						2004	+- 88	21
						2005	+- 88	21

TABLE FOUR (Continued)

EASTERN NORTH PACIFIC GRAY WHALE			NUMBER OF BELUGA WHALE IN COOK INLET			NUMBER OF BELUGA IN COOK INLET/0	
YEAR	NUMBER OF WHALES	RANK	YEAR	NUMBER OF WHALES	RANK	YEAR	NUMBER OF WHALES
1972	+-10,937.5	1	2002	0	1	2001	156.25
1971	+-12,500	2	2001	1.56	2	2018	212.5
1968	+-13,125	3	2006	+-12	3	2005	321.875
1969	+-14,687.5	4	1995	12.5	4	2006	325
1970	+-14,687.5	4	1996	28.125	5	2011	337.5
1975	+-15,000	5	1994	37.5	6	2016	337.5
1979	+-15,312.5	6	2018	37.5	6	1999	+-346,875
1993	+-15,625	7	2003	+-50	7	2010	350
2022	+-15,937.5	8	2000	+-62.5	8	2014	356.25
1973	+-16,250	9	1998	75	9	2003	375

TABLE FOUR (Continued)

EASTERN NORTH PACIFIC GRAY WHALES			NUMBER OF BELUGA WHALES IN COOK INLET			NUMBER OF BELUGA WHALES IN COOK INLET/0	
YEAR	NUMBER OF WHALES	RANK	YEAR	NUMBER OF WHALES	RANK	YEAR	NUMBER OF WHALES
2001	+-16,250	9	2011	75	9	2007	387.5
2002	+-16,562.5	10	2014	+-75	9	2012	400
1992	+-17,187.5	11	1997	100	10	2009	412.5
2003	+-17,187.5	11	2021	100	10	2002	412.5
2021	+-17,187.5	11	2012	112.5	11	2001	418.75
1974	+-17,500	12	2016	125	12	2004	437.5
2008	+-17,500	12	2010	+-125	12	1998	437.5
2000	+-17,500	12	2004	+-137.5	13	1997	450
1976	+-17,812.5	13	2008	+-187.5	14	2000	+-462.5
2004	+-18,125	14	2007	+-22.5	15	2008	475
1977	+-18,750	15	2005	+-231.25	16	1996	700
1999	+-19,062.5	16	1999	+-237	17	1995	737.5
2005	+-19,375	17	2009	+-266.7	18	1994	871

TABLE FOUR (Continued)

EASTERN NORTH PACIFIC GRAY WHALES (Continued)

YEAR	NUMBER OF WHALES	RANK
1978	+-19,687.5	18
1980	+-20,000	19
2020	+-20,000	19W

2009	+-20,000	19
2007	+-20,000	19
1991	+-20,000	19
2006	+-20,312.5	20
1994	+-20,321.5	21
1981	+-20,625	22
1995	+-20,625	22
2011	+-20,937	23
2010	+-21,250	24
1998	+-21,250	24
1996	+-21,250	24
1997	+-21,562.5	25
1990	21,875	26
1982	21,875	26
2019	+-22,187.5	27
1983	+-22,500	28

TABLE FOUR (Continued)

EASTERN NORTH PACIFIC GRAY WHALES (Continued)

YEAR	NUMBER OF WHALES	RANK
2018	+-22,500	28
1986	+-23,750	29
1984	+-23,750	30
1985	+-23,750	31

TABLE FOUR (Continued)

SOUTHERN RESIDENT KILLER WHALES (Continued)

YEAR	NUMBER OF WHALES	RANK
1968	+- 89	22
2011	+- 89	22
1998	+- 89	22
1990	+- 89	22
1968	+- 89	22
2006	90	23
1992	+-91	24
1964	+- 91	24
1997	+- 93	25
1991	+- 93	25
1965	+- 94	26
1966	+- 95	27
1994	+- 96	28
1967	+- 97	29
1996	+- 97	29
1993	+- 98	30
1995	+- 98	30

TABLE FOUR (Continued)

EASTERN NORTH PACIFIC GRAY WHALES (Continued)

YEAR	NUMBER OF WHALES	RANK
2012	+- 24,375	31
1989	+- 25,000	32
1987	+- 25,000	32
2017	+- 25,000	32
2016	+- 25,625	33
2013	+- 25,625	33
1988	+- 26,875	34
2015	+- 28,437	35
2014	+- 28,437	36

We constructed the tables above based on the line, bar graphs, or charts from the following sources:

- For the number of fin whales previously seen: (from chart to table).
  Source: Fin whales in the Northern Gulf of St Lawrence-Schleimer-2019-Ecology and Evolution-Wiley-Online library.
- For the number of Southern Resident Killer whales (from chart to table): NOAA Fisheries (Saving the Southern Resident Killer whales.
- For the number of Humpback whales in Glacier Bay & Icy Strait (chart to table): National Park & Reserve Alaska.
- For number of Eastern North Pacific Gray whale abundance 1967/68 - 2022/2023. Source: Gray Population Abundance/ NOAA Fisheries.

- For number of Beluga whales in Cook inlet: Beluga whale - Cook Inlet D.P.S (Delphinapterus leucas), (from chart to table). NOAA Fisheries Alaska Region Protected Resources Divisions Anchorage, Alaska and National Marine Fisheries Service. Alaska Fisheries Science Center-Marine Mammal Laboratory Seatle, Washington. August 12, 2022.

Regarding the number of beluga whales in Cook inlet/O: the base of the graph shows: Survey Year and number of days. This means that the survey was done in multiple days. So by putting for example 537.5/871 means records on different days.

- For North Atlantic Right whale Population (diagram to table): Downturn in Right whale population 1990-2017, NARWC
- For number of Bowhead whale (1) Relative abundance of bowheads (from graph to table): https://nammco.no/wp-content/uploads/2016/12 abundance-graph.
- For Ice Cover km2 (1): Idem
- For number of Blue whale population (from graph to table): Images for worldwide blue whale population ww.ontoha.org
- For Sei whale population size (Balaenoptera borealis) (from the graph-spaced 192019401960198020002020, and I had to figure out the in between year-to table): IUCN Red List
- For Narwal Population size (Female, Male) (from graph to table): NOAA Fisheries-National.
- For estimates number of North Atlantic Common Minke Whales (Sub-erea EB, EN, ES, EW, CM, CIC, CIP all included): Report of the Scientific Committee Virtual meeting, 25 April-13 May 2022. International Whaling Commission Cambridge UK, 2022

For table 1 Population estimates for Sperm whales used to make global population estimate: Current global population size post-whaling trend and historical trajectory of Sperm whales. Scientific reports.

- For table estimates Humpback whale at Bermuda, an Oceanic migratory stopover site: Frontiers-Managing Deep-Sea and Open Ocean Ecosystems at Ocean Basin scale - Volume 2.

TABLE FOUR (Continued): NUMBER OR ESTIMATES POPULATION WHALE, RANKING FROM THE LEAST TO THE GREATEST.

NORTH ATLANTIC RIGHT WHALE POPULATION			ESTIMATES OF NORTH ATLANTIC COMMON MINKE WHALE			POPULATION ESTIMATES FOR SPERM WHALES ANTARCTIC		
YEAR	NUMBER OF WHALES	RANK	YEAR	NUMBER OF WHALES	RANK	YEAR	NUMBER OF WHALES	RANK
1990	+-265.63	1	2004	6,246	1	2009-2010	22.4	1
1991	+-268.75	2	1993	8,371	2	2015-2016	380	2
1992	275	3	2010	10,991	3	2015	575	3
1993	+-278.12	4	1998	13,673	4	2003-2013	1842	4
1994	+-284.38	5	2019	15,693	5	2017-2018	1967	5
1995	+-290.63	6	1988	16,429	6	2014	2281	6
1996	+-306.25	7	2009	16,479	7	2016	4349	7
1999	+-307.82	8	1999	17,406	8	2017	5095	8

1997	+-309.38	9	2018	17,792	9	2005-2007	5445	9
1998	+-309.38	9	2003	19,377	10	2015	180	10

TABLE FOUR(Continued): NUMBER OR ESTIMATES OF POPULATION WHALE RANKING FROM THE LEAST TO THE GREATEST

NORTH ATLANTIC RIGHT WHALE POPULATION			ESTIMATES OF NORTH ATLANTIC COMMON MINKE WHALES			POPULATION ESTIMATES FOR SPERM WHALES: ANTARCTIC		
YEAR	NUMBER OF WHALES	RANK	YEAR	NUMBER OF WHALES	RANK	YEAR	NUMBER OF WHALES	RANK
2000	+-310.94	10	2011	21,218	11	2002-2007	13.557	11
2001	+-331.25	11	2014	23,059	12	1985-1991	17,500	12
2002	350	12	1996	23,522	13	1997	32,068	13
2003	359.38	13	2000	25,885	14	1986-1990	37,777	14
2004	+-371.88	14	1987	26,087	15	1982-1996	43,027	15
2005	+-392.2	15	1997	26,718	16	-	-	-
2006	+-403.13	16	2006	27,152	17	-	-	-
2017	411	17	2008	27,390	18	-	-	-
2007	+-415.63	18	2013	34,125	19	-	-	-
2016	+-434	19	2016	37,020	20	-	-	-
2008	437.5	20	2005	37,531	21	-	-	-
2015	+-453.13	21	2015	44,265	22	-	-	-

TABLE FOUR (Continued): NUMBER OR ESTIMATES OF POPULATION WHALE RANKING FROM THE LEAST TO THE GREATEST.

NORTH ATLANTIC RIGHT WHALE POPULATION			ESTIMATES OF NORTH ATLANTIC COMMON MINKE WHALES			POPULATION ESTIMATES FOR SPERM WHALES: ANTARCTIC		
YEAR	NUMBER OF WHALES	RANK	YEAR	NUMBER OF WHALES	RANK	YEAR	NUMBER OF WHALES	RANK
2009	+-459.38	22	2001	56,328	23	-	-	-
2014	462.5	23	2017	55,165	24	-	-	-
2013	+-471.89	24	1989	64,247	25	-	-	-
2012	475	25	2007	81,664	26	-	-	-
2011	+-476.56	26	1995	124,168	27	-	-	-
2010	+-478.13	27	-	-	-	-	-	-

TABLE FOUR (Continued): NUMBER OR ESTIMATES OF POPULATION WHALE, RANKING FROM THE LAST TO THE GREATEST

ESTIMATES HUMPBACK WHALE ABUNDANCE AT BERMUDA			NUMBER OF BOWHEAD WHALES (1)		
YEAR	NUMBER OF WHALES	RANK	YEAR	NUMBER OF WHALES	RANK
2020	71	1	1993	0,00	1
2010	78	2	1994	0,00	1

TABLE FOUR (Continued): NUMBER OR ESTIMATES OF POPULATION WHALE, RANKING FROM THE LAST TO THE GREATEST

ESTIMATES HUMPBACK WHALE ABUNDANCE AT BERMUDA			NUMBER OF BOWHEAD WHALE (1)		
YEAR	NUMBER OF WHALES	RANK	YEAR	NUMBER OF WHALES	RANK
2016	88	3	1999	0,00	1
2014	103	4	1996	0,0006	2
2011	104	5	1992	+-0,00125	3
2017	122	6	1983	+-0,00187.5	4
2018	148	7	1984	+-0,002188	5
2013	158	8	1981	+-0,0025	6
2012	165	9	1985	+-0,0025	6
2015	180	10	1987	+-0,0025	6
2019	218	11	1995	+-0,0025	6
-	-	-	1986	0,00281.3	7
-	-	-	1988	+-0,00281.3	7
-	-	-	1989	+-0,00375	8
-	-	-	1990	+-0,00375	8
-	-	-	1991	+-0,00625	9
-	-	-	1997	0,0075	10
-	-	-	1998	+-0,0088	11
-	-	-	2003	0,015	12
-	-	-	2001	0,015	12
-	-	-	2002	0,01	13

TABLE FOUR (Continued): NUMBER OR ESTIMATES OF POPULATION WHALE, RANKING FROM THE LEAST TO THE GREATEST

ESTIMATES HUMPBACK WHALE ABUNDANCE AT BERMUDA			NUMBER OF BOWHEAD WHALE (1)		
YEAR	NUMBER OF WHALES	RANK	YEAR	NUMBER OF WHALES	RANK
-	-	-	2005	+-0,2125	14
-	-	-	2004	+-0,025	15
-	-	-	2006	+-0,02625	16
-	-	-	2007	+-0,02625	16
-	-	-	2008	+-0,02625	16
-	-	-	2009	+-0,02625	16
-	-	-	2010	+-0,02688	17
-	-	-	2011	+-0,02688	17
-	-	-	2012	+-0,02688	17
-	-	-	2013	+-0,02688	17

TABLE 5: NUMBER OF POPULATION WHALE RANKING FROM THE LEAST TO THE GREATEST (Continued)

NUMBER OF BLUE WHALE POPULATION			SEI WHALE POPULATION SIZE		
YEAR	NUMBER OF WHALES	RANK	YEAR	NUMBER OF WHALES	RANK
1980	100	1	1990	+-23,750	1
1981	150	2	1991	+-24,687	2
1982	+-170	3	1992	25,000	3
1983	+-187.5	4	1993	25,000	3

1984	195	5	1994	+-25,078.13	4
1985	+-212.5	6	1995	+-26,000	5
1986	+-237,5	7	1996	+-26,000	5
1987	+-245	8	1997	+-27,500	6
1988	+-262.5	9	1998	+-27,578.13	7
1989	+-273	10	1999	+-29,998.33	8
1990	+-287.5	11	2000	+-29,998.34	9
1991	+-290	12	2001	+-29,998.40	10
1992	300	13	2002	30,000	11
1993	+-313	14	2003	31,000	12
1995	+-313	14	2004	+-32,000	13
1994	+-320	15	2005	+-32,000	13
1996	325	16	2006	+-32,500	14
1998	325	16	2007	+-32,600	15
1999	325	16	2008	+-33,750	16
1997	337.5	17	2009	+-33,900	17
2000	337.5	17	2010	+-34,500	18
2001	350	18	2011	35,000	19
2002	375	19	2012	+-35,750	20
2003	+-389	20	2013	+-35,800	21
2004	+-395	21	2014	+-37,450	22

NUMBER OF BLUE WHALE POPULATION			SEI WHALE POPULATION SIZE		
YEAR	NUMBER OF WHALES	RANK	YEAR	NUMBER OF WHALES	RANK
2005	+-408.3	22	2015	+-37,600	23
2006	425	23	2016	+-39,998.33	24
2007	+-437.5	24	2017	+-39,998.33	24
2008	+-455	25	2018	+-39,998.33	24
2009	475	26	2019	+-40,000	25
2010	+-487	28	2020	+-40,850	26
2011	+-480	27	-	-	-

TABLE 5 (Continued)

NARWHAL WHALE POPULATION SIZE (FEMALE) (07)			NARWHAL POPULATION WHALE SIZE (MALE) (07)		
YEAR	NUMBER OF WHALES	RANK	YEAR	NUMBER OF WHALES	RANK
1990	+-115.63	1	1990	+-115	1
1993	+-118.50	2	1991	+-149	2
1991	+-118.75	3	1992	+-154.17	3
1992	+-120.83	4	1993	+-155	4
1994	+-122.5	5	1994	+-160	5
1995	125	6	1995	162.5	6
2000	+-127	7	1996	+-165	7
1999	+-130	8	1997	+-172.5	8
1998	+-131.25	9	1998	+-173.5	9
1996	+-134.38	10	1999	175	10

1997	+-137.5	11	2000	+-177.08	11
2001	+-138	12	2001	+-189	12
2002	+-143.75	13	2002	+-197.5	13
2003	+-147.5	14	2020	+-207	14
2020	+-147.5	14	2022	+-208.33	15
2022	+-147.5	14	2003	+-209.38	16
2004	+-149.5	15	2021	212.5	17
2021	150	16	2019	+-218.75	18
2019	+-156.25	17	2005	225	19

TABLE 5 (Continued)

NARWHAL WHALE POPULATION SIZE (FEMALE) (07)			NARWHAL WHALE POPULATION SIZE (MALE) (07)		
YEAR	NUMBER OF WHALES	RANK	YEAR	NUMBER OF WHALES	RANK
2005	+-158.33	18	2006	+-235.42	20
2006	+-158.33	18	2007	+-247.5	21
2007	+-159.38	19	2017	250	22
2018	+-160	20	2008	+-256.25	23
2008	+-165	21	2016	+-262.5	24
2017	175	22	2009	+-270.83	25
2016	+-183.33	23	2015	+-272.5	26
2009	+-184.38	24	2012	+-274	27
2014	+-190	25	2013	275	28
2015	+-190	25	2010	+-277	29

2012	+-190.63	26	2011	+-277.5	30
2010	+-195.83	27	2014	+-277.5	30
2013	+-198	28	-	-	-

TABLE 6: ESTIMATES OF POPULATION OF WHALES<RANKING FROM THE LEAST TO THE GREATEST.

ESTIMATE NUMBER OF NORTH ATLANTIC COMMON MINKE WHALES		
YEAR	ESTIMATE NUMBER	RANK
2004	6,246	1
1993	8,371	2
2010	10,991	3
1998	13,673	4
2019	15,693	5
1988	16,429	6
2009	16,479	7
1999	17,406	8
2018	17,792	9
2003	19,377	10
2011	21,218	11
2014	23,059	12
1996	23,522	13
2000	25,885	14
1987	26,087	15
1997	26,718	16
2006	27,152	17

2008	27,390	18
2013	34,125	19

TABLE 6 (Continued)

ESTIMATE NUMBER OF NORTH ATLANTIC COMMON MINKE WHALES		
YEAR	ESTIMATE NUMBER	RANK
2016	37,020	20
2005	37,531	21
2015	44,265	22
2017	55,165	23
2001	56,328	24
1989	64,247	25
2007	81,664	26
1995	124,168	27

TABLE 6 (Continued)

POPULATION ESTIMATE FOR HUMPBACK WHALE ABUNDANCE AT BERMUDA			POPULATION ESTIMATE FOR SPERM WHALES. REGION: ANTARCTIC, CALIFORNIA			
YEAR	ESTIMATE NUMBER	RANK	YEAR	POPULATION ESTIMATE	RANK	REGION
2020	71	1	2009-2010	224	1	Canary Islands
2010	78	2	2015-2016	380	2	Off shore Ireland
2016	88	3	2015	575	3	Gulf of Alaska

POPULATION ESTIMATE FOR HUMPBACK WHALE ABUNDANCE AT BERMUDA			POPULATION ESTIMATE FOR SPERM WHALE. REGION: ANTARCTIC, CALIFORNIA			
YEAR	ESTIMATE NUMBER	RANK	YEAR	POPULATION ESTIMATE	RANK	REGION
2014	103	4	2003-2013	1842	4	Mediterranean Sea
2011	104	5	2017-2018	1967	5	Gulf of Mexico
2017	122	6	2014	2281	6	California Current
2018	148	7	2016	4349	7	North West Atlantic
2013	158	8	2017	5095	8	Haiwan Islands
2012	165	9	2005-2007	5445	9	Eastern North Atlantic
2015	180	10	2007	13,443	10	North Atlantic and Mediterranean
2019	218	11	2002-2007	13,557	11	North Atlantic and Northern-
			1985-1991	17,500	12	
			1997	32,068	13	
			1986-1990	37,777	14	
			1982-1996	43,027	15	

TABLE 7: NUMBER OF WHALES CAUGHT BY YEAR WITH ITS RANK (FROM THE LEAST TO THE GREATEST)

YEAR	NUMBER OF WHALES CAUGHT BY YEAR	RANK
1900	2,721	115
1901	2,416	116
1902	3,893	109
1903	4,233	103
1904	5,694	83
1905	5,356	87
1906	6,426	81
1907	7,804	80
1908	11,113	77
1909	12,876	72
1910	18,164	63

TABLE 7 (Continued)

YEAR	NUMBER OF WHALES CAUGHT BY YEAR	RANK
1911	25,064	52
1912	25,912	49
1913	25,700	50
1914	25,614	51
1915	22,523	54
1916	14,151	69
1917	10,193	79
1918	11,421	76

1919	14,240	68
1920	15,758	66
1921	12,098	74
1922	19,607	59
1923	18,472	61
1924	21,728	55
1925	29,330	45
1926	29,554	44
1927	28,268	47
1928	33,485	42
1929	42,303	35

TABLE 7 (Continued)

YEAR	NUMBER OF WHALES CAUGHT BY YEAR	RANK
1930	50,989	26
1931	18,211	62
1932	33,785	41
1933	36,644	38
1934	42,358	34
1935	41,619	36
1936	52,363	22
1937	66,569	12
1938	51,446	23
1939	45,736	32
1940	26,213	48
1941	16,391	65

YEAR	NUMBER OF WHALES CAUGHT BY YEAR	RANK
1942	12,894	71
1943	12,586	73
1944	11,665	75
1945	20,826	58
1946	35,555	40
1947	46,668	31
1948	48,570	29
1949	50,799	27
1950	51,038	25
1951	66,402	13
1952	51,402	24
1953	55,178	21
1954	67,882	10
1955	65,782	15
1956	65,923	14
1957	74,023	8
1958	78,951	6
1959	81,9997	4
1960	89,861	2
1961	62,306	17
1962	73,053	9
1963	83,052	3
1964	91,783	1
1965	79,805	5

1966	76,066	7
1967	67,628	11

TABLE 7 (Continued)

YEAR	NUMBER OF WHALES CAUGHT BY YEAR	RANK
1968	61,594	18
1969	62,805	16
1970	57,923	19
1971	57,651	20
1972	45,300	33
1973	49,485	28
1974	47,430	30
1975	39,500	37
1976	35,864	32
1977	30,172	43
1978	26,832	46
1979	24,093	53
1980	21,144	56
1981	20,897	57
1982	19,061	60
1983	16,532	64
1984	14,769	67
1985	13,430	70
1986	10,973	78

TABLE 7 (Continued)

YEAR	NUMBER OF WHALES CAUGHT BY YEAR	RANK
1987	5,797	82
1988	4,205	105
1989	4,388	102
1990	5,132	89
1991	3,782	111
1992	4,686	95
1993	4,221	104
1994	4,635	96
1995	3,582	113
1996	5,681	84
1997	5,557	85
1998	5,435	86
1999	4,926	90
2000	4,856	92
2001	5,141	88
2002	4,813	93
2003	4,866	91
2004	4,573	98
2005	4,476	99

TABLE 7 (Continued)

YEAR	NUMBER OF WHALES CAUGHT BY YEAR	RANK
2006	4,442	100
2007	4,594	97
2008	3,813	110
2009	4,096	106
2010	4,402	101
2011	3,952	107
2012	3,927	108
2013	4,807	94
2014	3,686	112
2015	3,094	114
2016	698	118
2017	1,274	117

TABLE 8: CARBON DIOXIDE EMISSIONS

YEAR	CO2 EMISSIONS
1900	0.10
1901	0.05
1902	0.04

TABLE 8 (Continued)

YEAR	CO2 EMISSIONS
1922	0.18
1923	0.46
1924	0.020
1925	0.05
1926	-0.07
1927	0.35
1928	-0.018
1929	0.33
1930	-0.34
1931	-0.47
1932	-0.35
1933	0.20
1934	0.30
1935	0.22
1936	0.40
1937	0.35
1938	-0.32
1939	0.25
1940	0.47

TABLE 8 (Continued)

YEAR	CO2 EMISSIONS
1941	0.15
1942	-0.025
1943	0.12
1944	0.11
1945	-0.8
1946	0.43
1947	0.55
1948	0.30
1949	-0.15
1950	0.79
1951	0.43
1952	0.08
1953	0.20
1954	0.15
1955	0.73
1956	0.50
1957	0.25
1958	0.20
1959	0.45

TABLE 8 (Continued)

YEAR	CO2 EMISSIONS
1960	0.60

1961	0.10
1962	0.40
1963	0.55
1964	0.65
1965	0.50
1966	0.60
1967	0.40
1968	0.75
1969	0.87
1970	1.20
1971	0.70
1972	0.74
1973	0.87
1974	-0.12
1975	0.05
1976	0.90
1977	0.55
1978	0.60

TABLE 8 (Continued)

YEAR	CO2 EMISSIONS
1979	0.58
1980	-0.20
1981	-0.50
1982	-0.22
1983	0.17

1984	0.73
1985	0.74
1986	0.38
1987	0.72
1988	0.85
1989	0.38
1990	0.40
1991	0.50
1992	-0.70
1993	0.23
1994	0.20
1995	0.50
1996	0.70
1997	0.20

TABLE 8 (Continued)

YEAR	CO2 EMISSIONS
1998	-0.15
1999	0.30
2000	0.70
2001	0.29
2002	0.65
2003	1.40
2004	1.38
2005	1
2006	1

2007	0.90
2008	0.60
2009	-0.49
2010	1.75
2011	1.15
2012	0.52
2013	0.33
2014	0.25
2015	-0.025
2016	-0.10

TABLE 8 (Continued)

YEAR	CO2 EMISSIONS
2017	0.45
2018	0.75
2019	-0.025
2020	1.60

TABLE 9: ANNUAL AVERAGE TEMPERATURE IN FAHRENHEIT WITH ITS RANK (FROM THE HIGHEST TO THE LOWEST)

YEAR	ANNUAL AVERAGE TEMPERATURE	RANK
1900	53.9	12
1901	53.5	16
1902	52.1	28
1903	50.6	43

1904	51.8	31
1905	51.7	32
1906	51.2	37
1907	52.4	25
1908	50.1	47
1909	51.8	31

TABLE 9 (Continued)

YEAR	ANNUAL AVERAGE TEMPERATURE	RANK
1910	54.5	4
1911	51.2	37
1912	50.9	40
1913	51.3	36
1914	53.2	19
1915	53.2	19
1916	51.1	38
1917	50.7	42
1918	52.3	26
1919	52.5	24
1920	51.3	36
1921	54.4	10
1922	51.2	37
1923	50.8	41
1924	51.5	34
1925	53.1	20

1926	53.9	12
1927	52.5	24
1928	50.7	42

TABLE 9 (Continued)

YEAR	ANNUAL AVERAGE TEMPERATURE	RANK
1929	49.3	52
1930	49.4	51
1931	49.5	50
1932	48.2	54.4
1933	50.5	44
1934	55.3	6
1935	51.9	30
1936	52.2	27
1937	51.4	35
1938	52.0	29
1939	51.7	32
1940	53.8	13
1941	51.5	34
1942	49.7	48
1943	52.2	27
1944	49.4	51
1945	50.5	44
1946	52.0	29
1947	51.6	33

TABLE 9 (Continued)

YEAR	ANNUAL AVERAGE TEMPERATURE	RANK
1948	50.6	43
1949	50.1	47
1950	51.6	33
1951	50.4	45
1952	51.1	38
1953	53.2	19
1954	52.8	22
1955	49.6	49
1956	51.5	34
1957	51.7	32
1958	53.6	15
1959	51.7	32
1960	51.9	30
1961	52.3	26
1962	50.6	43
1963	50.9	40
1964	48.3	53
1965	50.5	44
1966	52.4	25

TABLE 9 (Continued)

YEAR	ANNUAL AVERAGE TEMPERATURE	RANK
1967	51.9	30
1968	50.3	46
1969	52.0	29
1970	51.6	33
1971	50.5	44
1972	52.3	26
1973	51.4	35
1974	52.9	21
1975	50.8	41
1976	51.6	33
1977	53.2	19
1978	53.3	18
1979	52.7	23
1980	52.8	21
1981	54.3	11
1982	51.0	39
1983	53.5	16
1984	51.0	39
1985	51.9	30

TABLE 9 (Continued)

YEAR	ANNUAL AVERAGE TEMPERATURE	RANK
1986	52.7	23
1987	53.4	17
1988	53.3	18
1989	52.2	22
1990	53.4	17
1991	51.8	31
1992	53.6	15
1993	50.3	46
1994	54.7	8
1995	53.9	12
1996	54.3	11
1997	53.4	17
1998	52.9	21
1999	53.3	18
2000	53.7	14
2001	53.8	13
2002	52.0	29
2003	55.0	7
2004	52.1	28

YEAR	ANNUAL AVERAGE TEMPERATURE	RANK
2005	53.4	17
2006	53.8	13
2007	53.8	13
2008	51.9	30
2009	52.1	28
2010	52.7	23
2011	51.8	31
2012	56.6	1
2013	53.3	18
2014	55.6	5
2015	56.3	2
2016	56.2	3
2017	56.1	4
2018	56.2	-
2019	53.6	-
2020	55.7	-
2021	56.3	-
2022	55.9	-

TABLE 10: MAY TEMPERATURE ANOMALIES IN DEGREE CELSIUS (1901-2000 MEAN)

YEAR	MAY TEMPERATURE ANOMALIES IN DEGREE CELSIUS (1901-2000 Mean)	RANK
1900	- 0.05	86
1901	- 0.08	80

1902	- 0.27	20
1903	- 0.33	13
1904	- 0.45	4
1905	- 0.24	29
1906	- 0.20	38
1907	- 0.38	8
1908	- 0.34	9
1909	- 0.48	2
1910	- 0.30	17
1911	- 0.46	3
1912	- 0.18	45
1913	- 0.40	5
1914	- 0.19	44
1915	0.00	103
1916	- 0.25	27
1917	- 0.52	1
1918	- 0.39	7

TABLE 10 (Continued)

YEAR	MAY TEMPERATURE ANOMALIES IN DEGREE CELSIUS (1901-2000)	RANK
1919	- 0.22	34
1920	- 0.25	27
1921	- 0.21	35
1922	- 0.27	20
1923	- 0.27	20

1924	- 0.10	74
1925	- 0.24	29
1926	- 0.16	52
1927	- 0.20	38
1928	- 0.17	49
1929	- 0.31	14
1930	- 0.16	52
1931	- 0.13	63
1932	- 0.14	62
1933	- 0.24	29
1934	- 0.07	82
1935	- 0.23	33
1936	- 0.07	82
1937	- 0.01	98

TABLE 10 (Continued)

YEAR	MAY TEMPERATURE ANOMALIES IN DEGREE CELCIUS (1901-2000)	RANK
1938	- 0.03	92
1939	- 0.01	98
1940	0.16	122
1941	0.21	128
1942	0.15	121
1943	0.13	120
1944	0.24	132
1945	0.12	119

1946	- 0.02	96
1947	0.02	106
1948	0.05	113
1949	- 0.04	89
1950	- 0.02	96
1951	0.04	109
1952	0.08	115
1953	0.20	126
1954	- 0.12	65
1955	- 0.16	52
1956	- 0.27	20

TABLE 10 (Continued)

YEAR	MAY TEMPERATURE ANOMALIES IN DEGREE CELSIUS (1901-2000)	RANK
1957	0.05	113
1958	0.10	118
1959	0.04	109
1960	- 0.05	86
1961	0.04	109
1962	0.01	104
1963	- 0.02	96
1964	- 0.15	59
1965	- 0.20	38
1966	- 0.11	71
1967	0.21	128

1968	- 0.17	49
1969	0.18	125
1970	0.05	113
1971	- 0.03	92
1972	0.00	103
1973	0.24	132
1974	- 0.02	96
1975	0.08	115

TABLE 10 (Continued)

YEAR	MAY TEMPERATURE ANOMALIES IN DEGREE CELSIUS (1901-2000)	RANK
1976	- 0.12	65
1977	0.28	134
1978	0.10	118
1979	0.17	124
1980	0.38	146
1981	0.31	137
1982	0.17	124
1983	0.28	134
1984	0.31	137
1985	0.22	129
1986	0.35	143
1987	0.33	141
1988	0.48	150
1989	0.24	132

1990	0.50	151
1991	0.43	149
1992	0.37	144
1993	0.31	137
1994	0.32	139

TABLE 10 (Continued)

YEAR	MAY TEMPERATURE ANOMALIES IN DEGREE CELSIUS (1901-2000)	RANK
1995	0.32	139
1996	0.34	142
1997	0.41	147
1998	0.69	161
1999	0.33	141
2000	0.38	146
2001	0.58	155
2002	0.66	160
2003	0.63	157
2004	0.42	148
2005	0.70	162
2006	0.52	153
2007	0.65	158
2008	0.52	153
2009	0.62	156
2010	0.78	164
2011	0.55	154

2012	0.75	163
2013	0.66	160

TABLE 10 (Continued)

YEAR	MAY TEMPERATURE ANOMALIES IN DEGREE CELCIUS (1901-2000)	RANK
2014	0.87	169
2015	0.82	167
2016	0.99	174
2017	0.90	171
2018	0.81	166
2019	0.90	171
2020	0.99	174
2021	0.80	165
2022	0.86	168
2023	0.97	172

TABLE 11: CARBON DIOXIDE SINCE 1900 IN GIGATONS (Gt)

YEAR	CO2 EMISSIONS IN GIGATONS (Gt)
1900	0.125
1901	0.078125
1902	0.0625

TABLE 11: CARBON DIOXIDE SINCE 1900 IN GIGATONS (Gt) (Continued)

YEAR	CO2 EMISSIONS IN GIGATONS (Gt)
1903	0.1875
1904	0.025
1905	0.15
1906	0.125
1907	0.3125
1908	-0.125
1909	0.125
1910	0.15
1911	0.0625
1912	0.153
1913	0.28125
1914	-0.325
1915	-0.0625
1916	0.25
1917	0.153
1918	-0.0625
1919	-0.5
1920	0.5
1921	-0.45

TABLE 11: CO2 EMISSIONS SINCE 1900 IN GIGATONS (Gt) (Continued)

YEAR	CO2 EMISSIONS IN GIGATONS (Gt)
1922	0.153
1923	0.4375
1924	0.025
1925	0.0625
1926	-0.09375
1927	0.325
1928	-0.025
1929	0.275
1930	-0.325
1931	-0.4375
1932	-0.3125
1933	0.1873
1934	0.3123
1935	0.1875
1936	0.375
1937	0.3123
1938	-0.25
1939	0.25
1940	0.40625

TABLE 11: CO2 EMISSIONS SINCE 1900 IN GIGATONS (Gt) (Continued)

YEAR	CO2 EMISSIONS IN GIGATONS (Gt)
1941	0.09378
1942	-0.025
1943	0.0625
1944	0.0623
1945	-0.890625
1946	0.375
1947	0.5
1948	0.265625
1949	-0.187
950	0.75
1951	0.375
1952	0.09375
1953	0.1875
1954	0.125
1955	0.625
1956	0.5
1957	0.25
1958	0.21875
1959	0.4380

TABLE 11: CO2 EMISSIONS SINCE 1900 IN GIGATONS (Gt) (Continued)

YEAR	CO2 EMISSIONS IN GIGATONS (Gt)
1960	0.53125
1961	0.025
1962	0.3125
1963	0.525
1964	0.5625
1965	0.5
1966	0.541666
1967	0.375
1968	0.65
1969	0.875
1970	1.15
1971	0.6
1972	0.71875
1973	0.873
1974	-0.041666
1975	0.03125
1976	0.96875
1977	0.525
1978	0.583332

TABLE 11: CO2 EMISSIONS SINCE 1900 IN GIGATONS (Gt) (Continued)

YEAR	CO2 EMISSIONS IN GIGATONS (Gt)
1979	0.5625
1980	-0.125
1981	-0.4
1982	-0.66666
1983	-0.125
1984	0.65
1985	0.65625
1986	0.312
1987	0.625
1988	0.833332
1989	0.312
1990	0.35
1991	0.5
1992	-0.6875
1993	0.24
1994	0.15
1995	0.5
1996	0.688
1997	0.15

TABLE 11: CO_2 EMISSIOINS SINCE 1900 IN GIGATONS (Gt) (Continued)

YEAR	CO2 EMISSIONS IN GIGATONS (Gt)
1998	-0.083332
1999	0.333
2000	0.688
2001	0.25
2002	0.6
2003	1.35
2004	1.291666
2005	1.0
2006	1.0
2007	0.938
2008	0.5625
2009	-0.458332
2010	1.75
2011	1.15
2012	0.5
2013	0.3125
2014	0.25
2015	-0.03125
2016	-0.0625

TABLE 11: CO2 EMISSIONS SINCE 1900 IN GIGATONS (Gt) (Continued)

YEAR	CO2 EMISSIONS IN GIGATONS (Gt)
2017	0.46875
2018	0.75
2019	-0.023
2020	-1.59373
2021	1.6875

TABLE 12: TEN WARMEST YEARS (1880-2022)

RANK 1 = WARMEST PERIOD OF RECORD: 1880-2022	YEAR	ANOMALY IN DEGREE CELSIUS	ANOMALY IN FAHRENHEIT
1	2016	0.99	1.78
2	2020	0.98	1.76
3	2019	0.94	1.69
4	2015	0.93	1.67
5	2017	0.91	1.64
6	2022	0.86	1.55
7	2021	0.84	1.51

TABLE 12: TEN WARMEST YEARS (1880-2022) (Continued)

RANK 1 = WARMEST PERIOD OF RECORD: 1880-2022	YEAR	ANOMALY IN DEGREE CELSIUS	ANOMALY IN FAHRENHEIT
8	2018	0.82	1.48
9	2014	0.74	1.33
10	2010	0.72	1.30

Source: annual 2022 Global Climate Report/National Centers For Environmental Information (N.C.E.I)
Home/Climate Monitoring Report/ Annual 2022 Global Climate Report
June U.S. Release: Tue,11 Jul 2023, 11:00 AM EDT
https://www.ncei.noaa.gov/access/monitoring/monthly-report/global/202213

From the information of the ten warmest years (1880-2022) and the number of different types of whales in the tables above, using the whale's data available, we want to look at the correlation between the ten warmest years and the number of different types of whales.

Because of the difficulty to get the data on whale's population size, we take data of whales around 1960 to 2022. Based on that:

1. The year 2016, the first among the ten warmest years, has 85.72 Fin whales, ranking 21 out of 29 recorded from the least to the great number of fin whales. 24th out of 30 for Humpback whales. 16th out of 30 for Southern resident whales. 33rd out of 36 for Eastern north pacific gray whale. 12th out of 18 for Beluga whale in Cook inlet. 14th out of 20 for/0 whale. 19th out of 27 for North Atlantic Right whale. 20 out of 27 North Atlantic Common Mink whale. 7 out of 15 for Sperm whale (Atlantic California Current). 3rd out of 11 for estimates Humpback whale at Bermuda. No data recorded for relative abundance bowhead whale. No data recorded for blue whale, 24th out of 26 for Sei whale. 23rd out of 28 for Narwhal female population. 24th out of 30 for Narwal male population. 20th out of 27 for North Atlantic Common mink.
2. The year 2020, second among the ten warmest years, has no data recorded for fin whale, Humpback whale. 5th out of 30 for Southern Resident Killer whale. 19th out of 36 for Eastern north pacific Gray whale. No data recorded for Beluga whale in Cook inlet, for/0 whale. No data recorded for North Atlantic Right whale, North Atlantic Common Mink. No data recorded for Sperm whale Antarctic, California current. First out of 11 for estimates Humpback whale abundance at Bermuda. No data recorded for bowhead whale, Blue whale. 26th out of 26 for Sei. 14 out of 30 for male population Narwhal whale.
3. The year 2019, third among the ten warmest years, has no data recorded for Fin whale,Humpback whale, 6th out of 30 for Southern Resident Killer whale, 27th out of 36 for Eastern

North Atlantic Pacific Gray whale, no data recorded for Beluga whale in Cook inlet. No data recorded for/0, no data recorded for North Atlantic Right whale. 5th out of 27 for estimates of North Atlantic Common Minke whale. No data recorded for estimates Sperm whale in Antarctic, California Current. 11th out of 11 for estimates Humpback whale abundance at Bermuda. No data recorded for Blue whale, 25th out of 26 for Sei whale. 17th out of 28 for female population narwhal whale. 18th out of 30 for male narwhal whale population.

4. The year 2015, fourth among the ten warmest years, is 26th out of 29 of Fin whale. 24th out of 30 for Humpback whale. 14th out of 30 for Southern Resident Killer whale. 35th out of 36 for Eastern North Pacific Gray whale. No data recorded for Beluga whale in Cook inlet, no data for/0. 21rst out of 27 for North Atlantic Right whale. 22nd out of 27 for estimates of North Atlantic Common Minke whale. 3rd out of 15 for estimates Sperm whale in Antarctic, California Current. 10th out of 11 for estimates Humpback whale at Bermuda. No data recorded for Bowhead whale, Blue whale. 23rd out of 26 for Sei whale, 25th out of 28 for female Narwhal whale, 26th out of 30 for male Narwhal whale. 10th out of 11 for estimates Humpback whale at Bermuda.
5. The year 2017, fifth among the ten warmest years (1880-20220 has no data recorded for Fin whale, 18th out of 30 for Humpback whale, 10th out of 30 for Southern Resident Killer whale. 32nd out of 36 for Eastern North Pacific Gray whale. No data recorded for Beluga whale in Cook inlet, no data recorded for Beluga.../0. 17th out of 27 for North Atlantic Right whale, 24th out of 27 for estimates of North Atlantic Common Minke whale. 8th out of 15 for estimates for Sperm whale in Antarctic, California Current. 6th out of 11 for estimates Humpback whale at Bermuda. No data recorded for Bowhead whale, Blue whale. 24th out of 26 for Sei whale,

22nd out of 28 for female Narwhal whale. 22nd out of 30 for male Narwhal whale.

6. The year 2022, sixth among the ten warmest years (1880-2022), has no data recorded for Fin whale, Humpback whale, Southern Resident Killer, Eastern North Pacific Gray whale, Beluga whale in Cook inlet, for/0. No data for North Atlantic Right whale, estimates Sperm whale in Antarctic, California Current, estimates Humpback whale at Bermuda, Bowhead, Blue whale, Sei whale. 14th out of 28 for female Narwhal, 15th out of 30 for male Narwhal whale.
7. The year 2021, 7th among the ten warmest years (1880-2022), has no data recorded for Fin whale, Humpback whale. 7th out of 30 for Southern Resident Killer whale. 11th out of 36 for Eastern North Pacific Gray whale. 10th out of 18 for Beluga whale in Cook inlet, 1rst out of 20 for/0. No data recorded for North Atlantic Right whale. No data recorded for estimates North Atlantic Common minke whale. No data recorded for estimates Sperm whale, estimates Humpback at Bermuda, no data recorded for Bowhead whale, Blue whale, Sei whale. 16th out of 28 for female Narwhal whale, 17th out of 30 for male Narwhal whale.
8. The year 2018, 8th among the ten warmest years (1880-2022), has no data recorded for Fin whale, Humpback whale, 8th out of 30 for Southern Resident killer whale, 28th out of 36 for Eastern North Pacific Gray whale. 6th out of 18 for Beluga whale in Cook inlet, 2nd out of 20 for/0. No data recorded for North Atlantic Right whale. 9th out of 27 for estimates North Atlantic Common Minke whale. 5th out of 15 for estimates Sperm whale in Antarctic, California Current. 7th out of 11 for estimates Humpback whale at Bermuda. No data recorded for bowhead, Blue whale. 24th out of 26 for Sei whale, 20th out of 28 for female narwhal whale. No data for male Norwhal whale. No data for male Narwhal whale

9. The year 2014, 9th among the ten warmest years (1880-20220 is 16th out of 29 for Fin whale, 25th out of 30 for Humpback whale, 11th out of 30 for Southern Resident Killer whale, 36th out of 36 for Eastern Pacific Gray whale. 9th out of 18 for Beluga whale in Cook inlet; 8th out of 20 for .../ 0. 23rd out of 27 for North Atlantic Right whale, 12th out of 27 for estimates north Atlantic Common Mink whale. 6th out of 15 for estimates Sperm whale, 4th out of 11 for estimates humpback whale at Bermuda. No data recorded for Bowhead whale, Blue whale, 22nd out of 26 for Sei whale, 25th out of 28 for female Narwhal whale, 30th out of 30 for male Narwhal whale, 3rd out of 30 for male Narwhal whale.
10. The year 2010, tenth among the ten warmest years is 19th out of 29 for Fin whale. 27th out of 30 for Humpback whale, 19th out of 30 for Southern Resident Killer whale. 24th out of 36 for Eastern north Pacific Gray whale, 12th out of 18 for Beluga whale in Cook inlet. 7th out of 20 for.../0, 27th out of 27 for North Atlantic Right whale, 3rd out of 27 for estimates of North Atlantic Common Minke whale. 1rst out of 15 for estimates Sperm whale in Antarctic, California Current, 2nd out of 11 for estimates Humpback whale at Bermuda.

17th out of 17 for Bowhead whale, 28th out of 28 for Blue whale. 18th out of 26 for Sei whale, 27th out of 28 for female Narwhal whale. 29th out of 30 for male Narwhal whale.

After this , let us proceed with information in the following tables:

TABLE 13: NUMBER OF WHALES CAUGHT RANKING IN ORDER (FROM HIGHEST TO LOWEST), CO2 EMISSIONS IN GIGATONS

YEAR	NUMBER OF WHALES CAUGHT	RANK	YEAR	CO2 EMISSIONS IN GIGATONS	RANK
1964	91,783	1	1964	0.5625	19
1960	89, 861	2	1960	0.53125	21
1963	83.052	3	1963	0.525	22
1959	81,997	4	1959	0.4380	25
1965	79.805	5	1965	0.5	23
1958	78,951	6	1958	0.21875	40
1966	76,066	7	1966	0.541666	20
1957	74,023	8	1957	0.25	38
1962	73,053	9	1962	0.3125	32
1954	67,882	10	1954	0.125	45
1967	67,628	11	1967	0.375	28
1937	66,569	12	1937	0.3123	33
1951	66,402	13	1951	0.375	28

TABLE 13: NUMBER OF WHALES CAUGHT RANKING IN ORDER (HIGHEST TO LOWEST), CO2 EMISSIONS IN GIGATONS (Continued)

YEAR	NUMBER OF WHALES CAUGHT	RANK	YEAR	CO2 EMISSION IN GIGATONS	RANK
1956	65,923	14	1956	0.5	23
1955	65,782	15	1955	0.625	16

1969	62,805	16	1969	0.875	8
1961	62,306	17	1961	0.025	51
1968	61,594	18	1968	0.65	15
1970	57,923	19	1970	1.15	4
1971	57,651	20	1971	0.6	17
2001	55,178	21	2001	0.25	38
1936	52,363	22	1936	0.375	28
1938	51,446	23	1938	-0.25	61
1952	51,402	24	1952	0.09375	47
1950	51,038	25	1950	0.75	11
1930	50,989	26	1930	-0.325	62
1949	50,799	27	1949	-0.187	60
1973	49,485	28	1973	0.873	9
1948	48,570	29	1948	0.265625	37

TABLE 13: NUMBER OF WHALES CAUGHT RANKING IN ORDER (HIGHEST TO LOWEST), CO2 EMISSIONS IN GIGATONS. (Continued)

YEAR	NUMBER OF WHALES CAUGHT	RANK	YEAR	CO2 EMISSIONS IN GIGATONS	RANK
1974	47,430	30	1974	-0.41666	54
1947	46,668	31	1947	0.5	23
1939	45,736	32	1939	0.25	38
1972	45,300	33	1972	0.71875	12
1934	42,358	34	1934	0.3123	33
1929	42,303	35	1929	0.275	36

1935	41,619	36	1935	0.1875	41
1975	39,500	37	1975	0.03125	32
1933	36,644	38	1933	0.1873	42
1976	35,864	39	1976	0.96875	6
1946	35,555	40	1946	0.375	28

TABLE 13: NUMBER OF WHALES CAUGHT RANKING IN ORDER (FROM THE HIGHEST TO THE LOWEST), CO2 EMISSIONS IN GIGATONS (Continued)

YEAR	NUMBER OF WHALES CAUGHT	RANK	CO2 EMISSIONS IN GIGATONS	RANK
1932	33,785	41	- 0.3125	53
1928	33,485	42	-0.025	52
1977	30,172	43	0.525	22
1926	29,554	44	-0.09375	57
1925	29,330	45	0.0625	49
1978	26,832	46	0.583332	18
1927	28,268	47	0.325	31
1940	26,213	48	0.40625	27
1912	25,912	49	0.153	43
1913	25,700	50	0.28125	35
1914	25,614	51	-0.325	61
1911	25,064	52	0.0625	49
1979	24,093	53	0.5625	19
1915	22,523	54	-0.0625	54
1924	21,728	55	0.025	51
1980	21,114	56	-0.125	57

1981	20,897	57	-0.4	63
1945	20,826	58	-0.890625	67
1922	19,607	59	0.153	43

TABLE 13: NUMBER OF WHALES CAUGHT RANKING IN ORDER (FROM THE HIGHEST TO THE LOWEST), CO2 EMISSIONS IN GIGATONS (Continued)

1982	19,061	60	-0.166666	59
1923	18,472	61	0.4375	26
1931	18,211	62	-0.4375	64
1910	18,164	63	0.15	44
1983	16,532	64	-0.125	58
1941	16,391	65	0.09378	46
1920	15,758	66	0.5	23
1984	14,769	67	0.65	15
1919	14,240	68	-0.5	64
1916	14,151	69	0.25	38
1985	13,430	70	0.65625	14
1942	12,894	71	-0.025	52
1909	12,876	72	0.125	45
1943	12,586	73	0.0625	49
1921	12,098	74	-0.45	65
1944	11,665	75	0.0623	50
1918	11,421	76	-0.0625	54
1908	11,113	77	-0.125	57
1986	10,973	78	0.312	34

1917	10,193	79	0.153	43

TABLE 13: NUMBER OF WHALES CAUGHT RANKING IN ORDER (FROM THE HIGHEST TO THE LOWEST), CO2 EMISSIONS IN GIGATONS (Continued)

YEAR	NUMBER OF WHALES CAUGHT	RANK	YEAR	CO2 EMISSIONS IN GIGATONS	RANK
1907	7,804	80	1907	0.3125	32
1906	6,424	81	1906	0.125	45
1987	5,797	82	1987	0.625	16
1904	5,694	83	1904	0.025	51
1996	5,681	84	1996	0.688	13
1997	5,557	85	1997	0.15	44
1998	5,435	86	1998	-0.083332	56
1905	5,356	87	1905	0.15	44
2001	5,141	88	2001	0.25	38
1990	5,132	89	1990	0.35	29
1999	4,926	90	1999	0.333	30
2003	4,866	91	2003	1.35	2
2000	4,856	92	2000	0.688	13
2002	4,813	93	2002	0.6	17
2013	4,807	94	2013	0.3125	32
1992	4,686	95	1992	-0.6875	65
1994	4,635	96	1994	0.15	44
2007	4,594	97	2007	0.938	7
2004	4,573	98	2004	1.291666	3

TABLE 13: NUMBER OF WHALES CAUGHT RANKING IN ORDER (FROM HIGHEST TO LOWEST), CO2 EMISSIONS IN GIGATONS (Continued)

YEAR	NUMBER OF WHALES CAUGHT	RANK	YEAR	CO2 EMISSIONS IN GIGATONS	RANK
2005	4,476	99	2005	1.0	5
2006	4,442	100	2006	1.0	5
2010	4,402	101	2010	1.75	1
1989	4,388	102	1989	0.312	34
1903	4,233	103	1903	0.1875	41
1993	4,221	104	1993	0.24	39
1988	4,205	105	1988	0.8333.32	10
2009	4,096	106	2009	-0.4583.32	66
2011	3,952	107	2011	1.15	4
2012	3,927	108	2012	0.5	23
1902	3,893	109	1902	0.0625	49
2008	3,813	110	2008	0.5625	19
1991	3,782	111	1991	0.5	23
2014	3,686	112	2014	0.25	38
1995	3,582	113	1995	0.5	23

TABLE 13: NUMBER OF WHALES CAUGHT RANKING IN ORDER (FROM HIGHEST TO LOWEST), CO2 EMISSIONS IN GIGATONS (Continued)

YEAR	NUMBER OF WHALES CAUGHT	RANK	YEAR	CO2 EMISSIONS IN GIGATONS	RANK
2015	3,094	114	2015	-0.03125	53
1900	2,721	115	1900	0.125	45
1901	2,416	116	1901	0.078125	48
2017	1,274	117	2017	0.46875	24
2016	698	118	2016	-0.0625	55

TABLE 14: COMBINED TABLE OF NUMBER OF WHALES CAUGHT, CO2 EMISSIONS IN GIGATONS AND ANNUAL AVERAGE TEMPERATURE IN FAHRENHEIT

YEAR	NUMBER OF WHALES CAUGHT BY YEAR	RANK	YEAR	CO2 EMISSIONS IN GIGATONS BY YEAR	RANK	YEAR	ANNUAL AVERAGE TEMPE-RATURE IN FAHREN HEIT	RANK
1900	2,721	115	1900	0.125	45	1900	53.9	12
1901	2,416	116	1901	0.078125	48	1901	53.5	16
1902	3,893	109	1902	0.0625	49	1902	52.1	28
1903	4,233	103	1903	0.1875	41	1903	50.6	43

TABLE 14: COMBINED TABLE OF NUMBER OF WHALES CAUGHT, CO2 EMISSIONS IN GIGATONS AND ANNUAL AVERAGE TEMPERATURE IN FAHRENHEIT (Continued)

YEAR	NUMBER OF WHALES CAUGHT BY YEAR	RANK	YEAR	CO2 EMISSIONS IN GIGATONS BY YEAR	RANK	YEAR	ANNUAL AVERAGE TEMPE RATURE IN FAHREN HEIT	RANK
1904	5,694	83	1904	0.025	51	1904	51.8	31
1905	5,356	87	1905	0.15	44	1905	51.7	32
1906	6,424	81	1906	0.125	45	1906	51.2	37
1907	7,804	80	1907	0.3125	32	1907	52.4	25
1908	11,113	77	1908	-0.125	58	1908	50.1	47
1909	12,876	72	1909	0.125	45	1909	51.8	31
1910	18,164	63	1910	0.15	44	1910	54.5	9
1911	25,064	52	1911	0.0625	49	1911	51.2	37
1912	25,912	49	1912	0.153	43	1912	50.9	40
1913	25,700	50	1913	0.28125	35	1913	51.13	36
1914	25,614	51	1914	-0.325	61	1914	53.2	19
1915	22,523	54	1915	-0.0625	55	1915	53.2	19
1916	14,151	69	1916	0.25	38	1916	51.1	38
1917	10,193	79	1917	0.153	43	1917	50.7	42
1918	11,421	76	1918	-0.0625	55	1918	52.3	26

TABLE 14: COMBINED TABLE OF NUMBER OF WHALES CAUGHT, CO2 EMISSIONS INGIGATONS AND ANNUAL AVERAGE TEMPERATURE IN FAHRENHEIT (Continued)

YEAR	NUMBER OF WHALES CAUGHT BY YEAR	RANK	YEAR	CO2 EMISSIONS IN GIGATONS BY YEAR	RANK	YEAR	ANNUAL AVERAGE TEMPE-RATURE IN FAHREN HEIT	RANK
1919	14,240	68	1919	-0.5	67	1919	52.5	24
1920	15,758	66	1920	0.5	23	1920	51.3	36
1921	12,098	74	1921	-0.45	65	1921	54.4	10
1922	19,607	59	1922	0.153	43	1922	51.2	37
1923	18,472	61	1923	0.4375	26	1923	50.8	41
1924	21,728	55	1924	0.025	51	1924	51.5	34
1925	29,330	45	1925	0.0625	49	1925	53.1	20
1926	29,554	44	1926	-0.0937	57	1926	53.9	12
1927	28,268	46	1927	0.325	31	1927	52.5	24
1928	33,485	42	1928	-0.025	52	1928	50.7	42
1929	42,303	35	1929	0.275	36	1929	49.3	52
1930	50,989	26	1930	-0.325	62	1930	49.4	51
1931	18,211	62	1931	-0.4375	64	1931	49.5	50
1932	33,785	41	1932	-0.3125	53	1932	48.2	54
1933	36,644	38	1933	0.1873	42	1933	50.5	44

TABLE 14: COMBINED TABLE OF NUMBER OF WHALES CAUGHT, CO2 EMISSIONS IN GIGATONS AND ANNUAL AVERAGE TEMPERATURE IN FAHRENHEIT (Continued)

YEAR	NUMBER OF WHALES CAUGHT BY YEAR	RANK	YEAR	CO2 EMISSIONS IN GIGATONS BY YEAR	RANK	YEAR	ANNUAL AVERAGE TEMPE-RATURE IN FAHRENHEIT	RANK
1934	42,358	34	1934	0.3123	33	1934	55.3	6
1935	41,619	36	1935	0.1875	41	1935	51.9	30
1936	52.363	22	1936	0.375	28	1936	52.2	27
1937	66,569	12	1937	0.3123	33	1937	51.4	35
1938	51,446	23	1938	-0.25	61	1938	52.0	29
1939	45,736	32	1939	0.25	38	1939	51.7	32
1940	26,213	48	1940	0.40625	27	1940	53.8	13
1941	16,391	65	1941	0.09378	46	1941	51.5	34
1942	12,894	71	1942	-0.025	52	1942	49.7	48
1943	12,586	73	1943	0.0625	49	1943	52.2	27
1944	11,665	75	1944	0.0625	50	1944	49.4	51
1945	20,826	58	1945	-0.8906	69	1945	50.5	44
1946	35,555	40	1946	0.375	28	1946	52.0	29
1947	46,668	31	1947	0.25	38	1947	51.6	33

TABLE 14: COMBINED TABLE OF NUMBER OF WHALES CAUGHT BY YEAR, CO2 EMISSIONS IN GIGATONS BY YEAR AND ANNUAL AVERAGE TEMPERATURE IN FAHRENHEIT (Continued)

YEAR	NUMBER OF WHALES CAUGHT BY YEAR	RANK	YEAR	CO2 EMISSIONS IN GIGATONS BY YEAR	RANK	YEAR	ANNUAL AVERAGE TEMPERA-TURE IN FAHREN HEIT	RANK
1948	48,570	29	1948	0.26562	37	1948	50.6	43
1949	50,799	27	1949	-0.187	60	1949	50.1	47
1950	51,038	25	1950	0.75	11	1950	51.6	33
1951	66,402	13	1951	0.375	28	1951	50.4	45
1952	51,402	24	1952	0.09375	47	1952	51.1	38
1953	55,178	21	1953	0.1875	41	1953	53.2	19
1954	67,882	10	1954	0.125	45	1954	52.8	22
1955	65,782	15	1955	0.625	16	1955	49.6	49
1956	65,923	14	1956	0.5	23	1956	51.5	34
1957	74,023	8	1957	0.25	38	1957	51.7	32
1958	78,951	6	1958	0.21875	40	1958	53.6	15
1959	81,997	4	1959	0.3480	25	1959	51.7	32
1960	89,861	2	1960	0.53125	21	1960	51.9	30
1961	62,306	17	1961	0.025	51	1961	52.3	26

TABLE 14: COMBINED TABLE OF NUMBER OF WHALES CAUGHT BY YEAR, CO2 EMISSIONS IN GIGATONS BY YEAR AND ANNUAL AVERAGE TEMPERATURE IN FAHRENHEIT (Continued)

YEAR	NUMBER OF WHALES CAUGHT BY YEAR	RANK	YEAR	CO2 EMISSIONS IN GIGATONS BY YEAR	RANK	YEAR	ANNUAL AVERAGE TEMPE-RATURE IN FAHREN HEIT	RANK
1962	73,053	9	1962	0.3125	32	1962	50.6	43
1963	83,052	3	1963	0.525	22	1963	50.9	40
1964	91,783	1	1964	0.5625	19	1964	48.3	53
1965	79,805	5	1965	0.5	23	1965	50.5	44
1966	76,066	7	1966	0.54166	20	1966	52.4	25
1967	67,628	11	1967	0.375	28	1967	51.9	30
1968	61,594	18	1968	0.65	15	1968	50.3	46
1969	62,805	16	1969	0.875	8	1969	52.0	29
1970	57,923	19	1970	1.15	4	1970	51.6	33
1971	57,651	20	1971	0.6	17	1971	50.5	44
1972	45,300	33	1972	0.71875	12	1972	52.3	26
1973	49,485	28	1973	0.873	9	1973	51.4	35
1974	47,430	30	1974	-0.0416	54	1974	52.9	21
1975	39,500	37	1975	0.03125	32	1975	50.8	41

TABLE 14: COMBINED TABLE OF NUMBER OF WHALES CAUGHT BY YEAR, CO2 EMISSIONS IN GIGATONS BY YEAR AND ANNUAL AVERAGE TEMPERATURE IN FAHRENHEIT (Continued)

YEAR	NUMBER OF WHALES CAUGHT BY YEAR	RANK	YEAR	CO2 EMISSIONS IN GIGATONS BY YEAR	RANK	YEAR	ANNUAL AVERAGE TEMPE-RATURE IN FAHREN HEIT	RANK
1976	35,864	39	1976	0.96875	6	1976	51.6	33
1977	30,172	43	1977	0.525	22	1977	53.2	18
1978	26,832	47	1978	0.58333	18	1978	53.3	18
1979	24,093	53	1979	0.5625	19	1979	52.7	23
1980	21,144	56	1980	-0.125	58	1980	52.8	22
1981	20,897	57	1981	-0.4	63	1981	54.3	11
1982	19,061	60	1982	-0.1666	59	1982	51.0	39
1983	16,532	64	1983	-0.125	58	1983	53.5	16
1984	14,769	67	1984	0.65	15	1984	51.0	39
1985	13,430	70	1985	0.65625	14	1985	51.9	30
1986	10,973	78	1986	0.312	34	1986	52.7	23
1987	5,797	82	1987	0.625	16	1987	53.4	17
1988	4,205	105	1988	0.8333	10	1988	53.3	18
1989	4,388	102	1988	0.312	34	1989	52.2	27

TABLE 14: COMBINED TABLE OF NUMBER OF WHALES CAUGHT BY YEAR, CO2 EMISSIONS IN GIGATONS BY YEAR AND ANNUAL AVERAGE TEMPERATURE IN FAHRENHEIT (Continued)

YEAR	NUMBER OF WHALES CAUGHT BY YEAR	RANK	YEAR	CO2 EMISSIONS IN GIGATONS BY YEAR	RANK	YEAR	ANNUAL AVERAGE TEMPE RATURE IN FAHREN HEIT	RANK
1990	5,132	89	1990	0.35	29	1990	53.4	17
1991	3,782	111	1991	0.5	23	1991	51.8	31
1992	4,686	95	1992	-0.6875	68	1992	53.6	15
1993	4,221	104	1993	0.24	39	1993	50.3	46
1994	4,635	96	1994	0.15	44	1994	54.7	8
1995	3,582	113	1995	0.5	23	1995	53.9	12
1996	5,681	84	1996	0.688	13	1996	54,3	11
1997	5,557	85	1997	0.15	44	1997	53.4	17
1998	5,435	86	1998	-0.0833	56	1998	52.9	21
1999	4,926	90	1999	0.333	30	1999	53.3	18
2000	4,856	92	2000	0.688	13	2000	53.7	14
2001	5,141	88	2001	0.25	38	2001	53.8	13
2002	4,813	93	2002	0.6	17	2002	52.0	29
2003	4,866	91	2003	1.35	2	2003	55.0	7

TABLE 14: COMBINED TABLE OF NUMBER OF WHALES CAUGHT BY YEAR, CO2 EMISSIONS IN GIGATONS BY YEAR AND ANNUAL AVERAGE TEMPERATURE IN FAHRENHEIT (Continued)

YEAR	NUMBER OF WHALES CAUGHT BY YEAR	RANK	YEAR	CO2 EMISSIONS IN GIGATONS BY YEAR	RANK .	YEAR	ANNUAL AVERAGE TEMPE RATURE IN FAHREN HEIT	RANK
2004	4,573	98	2004	1.29166	3	2004	52.1	28
2005	4,476	99	2005	1.0	5	2005	53.4	17
2006	4,442	100	2006	1.0	5	2006	53.8	13
2007	4,594	97	2007	0.938	7	2007	53.8	13
2008	3,813	110	2008	0.5625	19	2008	51.9	30
2009	4,096	106	2009	-0.4583	66	2009	52.1	28
2010	4,402	101	2010	1.75	1	2010	52.7	23
2011	3,952	107	2011	1.15	4	2011	51.8	31
2012	3,927	108	2012	0.5	23	2012	56.6	1
2013	4,807	94	2013	0.3125	32	2013	53.3	18
2014	3,686	112	2014	0.25	38	2014	55.6	5
2015	3,094	114	2015	-0.3125	53	2015	56.3	2
2016	698	118	2016	-0.0625	55	2016	56.2	3
2017	1,274	117	2017	0.46875	24	2017	56.1	4

To provide a complementary views, we have a line graph on whales caught here bellow.

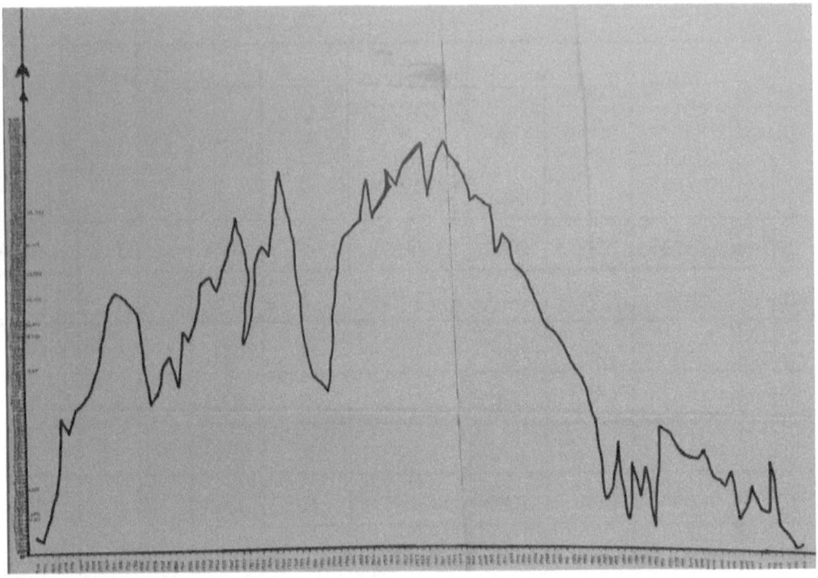

After the line graph above, let us proceed with the tables.

TABLE 15: COMBINED TABLE OF NUMBER OF WHALES CAUGHT, CO2 EMISSIONS IN GIGATONS AND ANNUAL AVERAGE TEMPERATURE IN FAHRENHEIT, ALL RANKED IN ORDER

YEAR	NUMBER OF WHALES CAUGHT	RANK	YEAR	ANNUAL AVERAGE TEMPE RATURE IN FAHREN-HEIT	RANK	YEAR	CO2 EMISSIONS IN GIGATONS	RANK
1964	91,783	1	2012	56.6	1	2010	1.75	1
1960	89,861	2	2015	56.3	2	2003	1.35	2
1963	83,052	3	2016	56.2	3	2004	1.29166	3
1959	81,997	4	2017	56.1	4	1970	1.15	4
1965	79,805	5	2014	55.6	5	2011	1.15	4
1958	78,951	6	1934	55.3	6	2005	1.0	5
1966	76,066	7	2003	55.0	7	2006	1.0	5
1957	74,023	8	1994	54.7	8	1976	0.96875	6
1962	73,053	9	1910	54.5	9	2007	0.938	7
1954	67,882	10	1921	54.4	10	1969	0.875	8
1967	67,628	11	1996	54.3	11	1973	0.873	9
1937	66,569	12	1995	53.9	12	1988	0.8333	10
1951	66,402	13	2001	53.8	13	1950	0.75	11

TABLE 15: COMBINED TABLE OF NUMBER OF WHALES CAUGHT, ANNUAL AVERAGE TEMPERATURE IN FAHRENHEIT AND CO2 EMISSIONS IN GIGATONS (Continued)

YEAR	NUMBER OF WHALES CAUGHT	RANK	YEAR	ANNUAL AVERAGE TEMPE-RATURE IN FAHREN-HEIT	RANK	YEAR	CO2 EMISSIONS IN GIGATONS	RANK
1956	65,923	14	2000	53.7	14	1972	0.71875	12
1955	65,782	15	1992	53.6	15	2000	0.688	13
1969	62,805	16	2006	53,8	13	1985	0.65625	14
1961	62,306	17	1940	53.8	13	1984	0.65	15
1968	61,594	18	2007	53.8	13	1968	0.65	15
1970	57,923	19	1958	53.6	15	1987	0.625	16
1971	57,651	20	1983	53.5	16	1955	0.625	16
1953	55,178	21	1901	53.5	16	2002	0.6	17
1936	52.363	22	1983	53.5	16	1971	0.6	17
1938	51,446	23	2005	53.4	17	1978	0.5833	18
1952	51,402	24	1997	53.4	17	1979	0.5625	19
1950	51,038	25	1987	53.4	17	1964	0.5625	19
1930	50,989	26	1990	53.4	17	2008	0.5625	19
1949	50,799	27	2013	53.3	18	1966	0.54166	20
1973	49,485	28	1999	53.3	18	1960	0.53125	21

TABLE 15: COMBINED TABLE OF NUMBER OF WHALES CAUGHT, CO2 EMISSIONS IN GIGATONS, ANNUAL AVERAGE TEMPERATURE IN FAHRENHEIT, ALL RANKED IN ORDER (Continued)

YEAR	NUMBER OF WHALES CAUGHT	RANK	YEAR	ANNUAL AVERAGE TEMPE-RATURE	RANK	YEAR	CO2 EMISSIONS IN GIGATONS	RANK
1948	48,570	29	1978	53,3	18	1977	0.525	22
1974	47,430	30	1988	53.3	18	1920	0.5	23
1947	46,668	31	1953	53.2	19	1956	0.5	23
1939	45,736	32	1977	53.2	19	1963	0.525	22
1972	45,300	33	1914	53.2	19	1995	0.5	23
1934	42,358	34	1915	53.2	19	1991	0.5	23
1929	42,303	35	1925	53.1	20	1965	0,5	23
1935	41,619	36	1974	52.9	21	-	-	-
1975	39,500	37	1998	52.9	21	2012	0.5	23
1933	36,644	38	1980	52.8	22	1920	0.5	23
1976	35,864	39	1954	52.8	22	2017	0.46875	24
1946	35,555	40	1986	52.7	23	1959	0.4380	25
1932	33,785	41	1979	52.7	23	1923	0.4375	26
1928	33,485	42	2010	52.7	23	1940	0.40625	27
977	30,172	43	1919	52.5	24	1936	0.375	28

TABLE 15: COMBINED TABLE OF NUMBER OF WHALES CAUGHT, CO2 EMISSIONS IN GIGATONS AND ANNUAL AVERAGE TEMPERATURE IN FAHRENHEIT, ALL RANKED IN ORDER (Continued)

YEAR	NUMBER OF WHALES CAUGHT	RANK	YEAR	ANNUAL AVERAGE TEMPE-RATURE	RANK	YEAR	CO2 EMISSIONS IN GIGATONS	RANK
1926	29,554	44	1966	52.4	25	1946	0.375	28
1925	29,330	45	1927	52.5	24	1951	0.375	28
1978	26,832	47	1907	52.4	25	1967	0.375	28
1927	28,268	46	1972	52.3	26	1990	0.35	29
1940	26,213	48	1961	52.3	26	1999	0.333	30
1912	25,912	49	1918	52.3	26	1927	0.325	31
1913	25,700	50	1936	52.2	27	1975	0.03125	32
1914	25,614	51	1943	52.2	27	1937	0.3125	33
1911	25,064	52	1989	52.2	27	1962	0.3125	32
1979	24,093	53	2009	52.1	28	1907	0.3125	32
1915	22,523	54	2004	52.1	28	1934	0.3125	33
1924	21,728	55	1902	52.1	28	2013	0.3125	32
1980	21,144	56	1969	52.0	29	1989	0.312	34
1981	20,897	57	1946	52.0	29	1986	0.312	34
1945	20,826	58	1938	52.0	29	1913	0.28125	35

TABLE 15: COMBINED TABLE OF NUMBER OF WHALES CAUGHT, CO2 EMISSIONS IN GIGATONS AND ANNUAL AVERAGE TEMPERATURE IN FAHRENHEIT, ALL RANKED IN ORDER (Continued)

YEAR	NUMBER OF WHALES CAUGHT	RANK	YEAR	ANNUAL AVERAGE TEMPE-RATURE	RANK	YEAR	CO2 EMISSIONS IN GIGATONS	RANK
1922	19,607	59	2002	52.0	29	1929	0.275	36
1982	19,061	60	2008	51.9	30	1948	0.26562	37
1923	18,472	61	1935	51.9	30	1939	0.25	38
1931	18,211	62	1967	51.9	30	1947	0.25	38
1910	18,164	63	1960	51.9	30	1957	0.25	38
1983	16,532	64	1985	51.9	30	2001	0.25	38
1941	16,391	65	2011	51.8	31	2014	0.25	38
1920	15,758	66	1991	51.8	31	1916	0.25	38
1984	14,769	67	1909	51.8	31	1993	0.24	39
1919	14,240	68	1904	51.8	31	1958	0.21875	40
1916	14,151	69	1959	51.7	32	1903	0.1875	41
1985	13,430	70	1957	51.7	32	1935	0.1875	41
1942	12,894	71	1939	51.7	32	1953	0.1875	41
1909	12,876	72	1905	51.7	32	1933	0.1873	42
1943	12,586	73	1976	51.6	33	1917	0.153	43

TABLE 15: COMBINED TABLE OF NUMBER OF WHALES CAUGHT, CO2 EMISSIONS IN GIGATONS AND ANNUAL AVERAGE TEMPERATURE IN FAHRENHEIT, ALL RANKED IN ORDER (Continued)

YEAR	NUMBER OF WHALES CAUGHT	RANK	YEAR	ANNUAL AVERAGE TEMPE-RATURE	RANK	YEAR	CO2 EMISSIONS IN GIGATONS	RANK
1921	12,098	74	1947	51.6	33	1912	0.153	43
1944	11,665	75	1950	51.6	33	1922	0.153	43
1918	11,421	76	1970	51.6	33	1905	0.15	44
1908	11,113	77	1956	51.5	34	1910	0.15	44
1986	10,973	78	1941	51.5	34	1994	0.15	44
1917	10,193	79	1973	51.4	35	1997	0.15	44
1907	7,804	80	1937	51.4	35	1900	0.125	45
1906	6,424	81	1920	51.3	36	1906	0.125	45
1987	5,797	82	1913	51.3	36	1909	0.125	45
1904	5,694	83	1922	51.2	37	1954	0.125	45
1996	5,681	84	1911	51.2	37	1901	0.07812	48
1997	5,557	85	1906	51.2	37	1952	0.09375	47
1998	5,435	86	1952	51.1	38	1941	0.09378	46
1905	5,356	87	1916	51.1	38	1943	0.0625	49
2001	5,141	88	1982	51.0	39	1902	0.0625	49

TABLE 15: COMBINED TABLE OF NUMBER OF WHALES CAUGHT, CO2 EMISSIONS IN GIGATONS AND ANNUAL AVERAGE TEMPERATURE IN FAHRENHEIT, ALL RANKED IN ORDER (Continued)

YEAR	NUMBER OF WHALES CAUGHT	RANK	YEAR	ANNUAL AVERAGE TEMPE-RATURE	RANK	YEAR	CO2 EMISSIONS IN GIGATONS	RANK
1989	4,388	89	1963	50.9	40	1925	0.0625	49
1999	4,926	89	1912	50.9	40	1911	0.0625	49
2003	4,866	90	1975	50.8	41	1944	0.0623	50
2000	4,856	91	1923	50.8	41	1961	0.025	51
2002	4,813	92	1928	50.7	42	1904	0.025	51
2013	4,807	93	1917	50.7	42	1924	0.025	51
1992	4,686	94	1962	58.6	43	1942	-0.025	52
1994	4,635	95	1948	50.6	43	1928	-0.025	52
2007	4,594	96	1903	50.6	43	1932	-0.3125	53
2004	4,573	97	1971	50.5	44	2015	-0.3125	53
2005	4,476	98	1965	50.5	44	1915	-0.0625	55
2006	4,442	99	1945	50.5	44	1918	-0.0625	55
2010	4,402	100	1933	50.5	44	1974	-0.0416	54
1989	4,388	101	1951	50.4	45	2016	-0.0625	55
1903	4,233	102	1968	50.3	46	1998	-0.0833	56

TABLE 15: COMBINED TABLE OF NUMBER OF WHALES CAUGHT, CO2 EMISSIONS IN GIGATONS AND ANNUAL AVERAGE TEMPERATURE IN FAHRENHEIT, ALL RANKED IN ORDER (Continued)

YEAR	NUMBER OF WHALES CAUGHT	RANK	YEAR	ANNUAL AVERAGE TEMPE-RATURE	RANK	YEAR	CO2 EMISSIONS IN GIGATONS	RANK
1993	4,221	103	1993	50.3	46	1926	-0.0937	57
1988	4,205	104	1908	50.1	47	1908	-0.125	58
2009	4,096	105	1949	50.1	47	1983	-0.125	58
2011	3,952	106	1942	49.7	48	1980	-0125	58
2012	3,927	107	1955	49.6	49	1982	-0.1666	59
1902	3,893	108	1931	49.5	50	1949	-0.187	60
2008	3,813	109	1944	49.4	51	1938	-0.25	61
1991	3,782	110	1930	49.4	51	1914	-0.325	62
2014	3,686	111	1929	49.3	52	1930	-0.325	62
1995	3,582	112	1964	48.3	53	1981	-0.4	63
2015	3,094	113	1932	48.2	54	1919	-0.5	67
1900	2,721	114	-	-	-	1931	-0.4375	64
1901	2,416	115	1926	53.9	12	1921	-0.45	65
2017	1,274	116	1924	51.5	34	2009	-045833	66

TABLE 15: COMBINED TABLE OF NUMBER OF WHALES CAUGHT, CO2 EMISSIONS IN GIGATONS AND ANNUAL AVERAGE TEMPERATURE IN FAHRENHEIT, ALL RANKED IN ORDER (Continued)

YEAR	NUMBER OF WHALES CAUGHT	RANK	YEAR	ANNUAL AVERAGE TEMPE-RATURE	RANK	YEAR	CO2 EMISSIONS IN GIGATONS	RANK
2016	698	117	1981	54.3	11	1992	-0.6875	68
-	-	-	1984	51.0	39	1945	-0.8906	69
-	-	-	1900	53.9	12	1996	0.688	13

Since the rise of temperature is caused mainly by the excess of the carbon dioxide released in the atmosphere, while the natural sinks including whales help to absorb, reduce the CO2's excess; we will use the data in the tables above to spot out the influence between the number of whales caught, the CO2 emissions and the rise of temperature (referring to the annual average temperatures). Based on that:

1. The year 1964 ranking first in number of whales caught, with a CO2 emissions of 0.5625 gigatons ranking 19th out of 69, has an annual average temperature of 48.3 Fahrenheit ranking 53rd out of 54
2. The year 1960 ranking second in number of whales caught, with a CO2 emissions of 0.53125 gigatons ranking 21st out of 69, has an annual average temperature of 51.9 Fahrenheit ranking 30th out of 54.
3. The year 1963 ranking 3rd in number of whales caught, with a CO2 emissions of 0.525 gigatons ranking 22nd out of 69, has an annual average of 50.9 Fahrenheit ranking 40th out of 54.
4. The year 1959 ranking 4th in number of whales caught, with a Co2 emissions of 0.4380 gigatons ranking 25th out of 69,

has an annual average temperature of 51.7 Fahrenheit ranking 32nd out of 54.

5. The year 1965 ranking 5th in number of whales caught, with a CO_2 emissions of 0.5 gigatons ranking 23rd out of 69, has an annual average temperature of 50.5 Fahrenheit ranking 44th out of 54.
6. The year 1958 ranking 6th in number of whales caught, with a CO_2 emission of 0.21875 gigatons ranking 40th out of 69, has an annual average temperature of 53.8 Fahrenheit ranking 15th out of 54.
7. The year 1966 ranking 7th in number of whales caught, with a CO_2 emission of 0.541666 gigatons ranking 20th out of 69, has an annual average temperature of 52.4 Fahrenheit ranking 25th out of 54.
8. The year 1957 ranking 8th in number of whales caught, with a CO_2 emission of 0.25 gigatons ranking 38th out of 69, has an annual average temperature of 51.7 Fahrenheit ranking 32nd out of 54.
9. The year 1962 ranking 9th in number of whales caught, with a CO_2 emission of 0.3125 gigatons ranking 32nd out of 69, has an annual average temperature of 50.6 Fahrenheit ranking 43rd out of 54.
10. The year 1954 ranking 10th in number of whales caught, with a CO_2 emission of 0.125 gigatons ranking 45th out of 69, has an annual average temperature of 52.8 Fahrenheit ranking 22nd out of 54.
11. The year 1967 ranking 11th in number of whales caught, with a CO_2 emission of 0.375 gigatons ranking 28th out of 69, has an annual average temperature of 51.9 Fahrenheit ranking 30th out of 54.
12. The year 1937 ranking 12th in number of whales caught, with a CO_2 emission of 0.3123 gigatons ranking 33rd out of 69, has an annual average temperature of 51.4 Fahrenheit ranking 35th out of 54.

13. The year 1951 ranking 13th in number of whales caught, with a CO2 emission of 0.375 gigatons ranking 28th out of 69, has an annual average temperature of 50.4 Fahrenheit ranking 45th out of 54.
14. The year 1956 ranking 14th in number of whales caught, with a CO2 emission of 0.5 gigatons ranking 23rd out of 69, has an annual average temperature of 51.5 Fahrenheit ranking 34th out of 54.
15. The year 1955 ranking 15th in number of whales caught, with a CO2 emission of 0.625 gigatons ranking 16th out of 69, has an annual average temperature of 49.6 fahrenheit ranking 49th out of 54.
16. The year 1969 ranking 16th in number of whales caught, with a CO2 emission of 0.875 gigatons ranking 8th out of 69, has an annual average temperature of 52.0 Fahrenheit ranking 29th out of 54.
17. The year 1961 ranking 17th in number of whales caught, with a CO2 emission of 0.025 gigatons ranking 51st out of 69, has an annual average temperature of 52.3 Fahrenheit ranking 26th out of 54.
18. The year 1968 ranking 18th in number of whales caught, with a CO2 emission of 0.65 gigatons ranking 15th out of 69, has an annual average temperature of 50.3 Fahrenheit ranking 46th out of 54.
19. The year 1970 ranking 19th in number of whales caught, with a CO2 emission of 1.15 gigatons ranking 4th out of 69, has an annual average temperature of 51.6 Fahrenheit ranking 33rd out of 54.
20. The year 1971 ranking 20th in number of whales caught, with a CO2 emission of 0.6 gigatons ranking 17th out of 69, has an annual average temperature of 50.5 Fahrenheit ranking 44th out of 54.
21. The year 1953 ranking 21st in number of whales caught, with a CO2 emission of 0.1875 gigatons ranking 41st out of 69,

has an annual average temperature of 53.2 Fahrenheit ranking 19th out of 54.

22. The year 1936 ranking 22nd in number of whales caught, with a CO_2 emission of 0.375 gigatons ranking 28th out of 69, has an annual average temperature of 52.2 Fahrenheit ranking 27th out of 54.
23. The year 1938 ranking 23rd in number of whales caught, with a CO_2 emission of -0.25 gigatons ranking 61st out of 69, has an annual average temperature of 52.0 Fahrenheit ranking 29th out of 54.
24. The year 1952 ranking 24th in number of whales caught, with a CO_2 emission of 0.09375 gigatons ranking 47th out of 69, has an annual average temperature of 51.1 ranking 38th out of 54.
25. The year 1950 ranking 25th in number of whales caught, with a CO_2 emission of 0.75 gigatons ranking 11th out of 69, has an annual average temperature of 51.6 Fahrenheit ranking 33rd out of 54.
26. The year 1930 ranking 26th in number of whales caught, with a CO_2 emission of -0.325 gigatons ranking 62nd out of 69, has an annual average temperature of 49.4 Fahrenheit ranking 51st out of 54.
27. The year 1949 ranking 27th in number of whales caught, with a CO_2 emission of -0.187 gigatons ranking 60th out of 69, has an annual average temperature of 50.1 Fahrenheit ranking 47th out of 54.
28. The year 1973 ranking 28th in number of whales caught, with a CO_2 emission of 0.873 gigatons ranking 9th out of 69, has an annual average temperature of 51.4 Fahrenheit ranking 35th out of 54.
29. The year 1948 ranking 29th in number of whales caught, with a CO_2 emission of 0.265625 gigatons ranking 54th out of 69, has an annual average temperature of 50.6 Fahrenheit ranking 43rd out of 54.

30. The year 1974 ranking 30th in number of whales caught, with a CO_2 emission of -0.041666 gigatons ranking 54th out of 69, has an annual average temperature of 52.9 Fahrenheit ranking 21st out of 54.
31. The year 1947 ranking 31st in number of whales caught, with a CO_2 emission of 0.25 gigatons ranking 38th out of 69, has an annual average temperature of 51.6 Fahrenheit ranking 33rd out of 54.
32. The year 1939 ranking 32nd in number of whales caught, with a CO_2 emission of 0.25 gigatons ranking 38th out of 69, has an annual average temperature of 51.7 Fahrenheit ranking 32nd out of 54.
33. The year 1972 ranking 33rd in number of whales caught, with a CO_2 emission of 0.71875 gigatons ranking 12th out of 69, has an annual average temperature of 52.3 Fahrenheit ranking 26th out of 54.
34. The year 1934 ranking 34th in number of whales caught, with a CO_2 emission of 0.3123 gigatons ranking 33rd out of 69, has an annual average temperature of 55.3 Fahrenheit ranking 6th out of 54.
35. The year 1929 ranking 35th in number of whales caught, with a CO_2 emission of 0.275 gigatons ranking 36th out of 69, has an annual average temperature of 49.3 Fahrenheit ranking 52nd out of 54.
36. The year 1935 ranking 36th in number of whales caught, with a CO_2 emission of 0.1875 gigatons ranking 41st out of 69, has an annual average temperature of 51.9 Fahrenheit ranking 30th out of 54.
37. The year 1975 ranking 37th in number of whales caught, with a CO_2 emission of 0.03125 gigatons ranking 32nd out of 69, has an annual average temperature of 50.8 Fahrenheit ranking 41st out of 54.
38. The year 1933 ranking 38th in number of whales caught, with a CO_2 emission of 0.1873 gigatons ranking 42nd out of 69,

has an annual average temperature of 50.5 Fahrenheit ranking 44th out of 54.

39. The year 1976 ranking 39th in number of whales caught, with a CO2 emission of 0.96875 gigatons ranking 6th out of 69, has an annual average temperature of 51.6 Fahrenheit ranking 33rd out of 54.
40. The year 1946 ranking 40th in number of whales caught, with a CO2 emission of 0.375 gigatons ranking 28th out of 69, has an annual average temperature of 52.0 Fahrenheit ranking 29th out of 54.
41. The year 1932 ranking 41st in number of whales caught, with a CO2 emission of -03125 gigatons ranking 53rd out fo 69, has an annual average temperature of 48.2 Fahrenheit ranking 54th out of 54.
42. The year 1928 ranking 42nd in number of whales caught, with a CO2 emission of -0.025 gigatons ranking 52nd out of 69, has an annual average temperature of 50.7 Fahrenheit ranking 42nd out of 54.
43. The year 1977 ranking 43rd in number of whales caught, with a CO2 emission of 0.525 gigatons ranking 22nd out of 69, has an annual average temperature of 53.2 Fahrenheit ranking 19th out of 54.
44. The year 1926 ranking 44th in number of whales caught, with a CO2 emission of -0.09375 gigatons ranking 57th out of 69, has an annual average temperature of 53.9 Fahrenheit ranking 12th out of 54.
45. The year 1925 ranking 45th in number of whales caught, with a CO2 emission of 0.0625 gigatons ranking 49th out of 69, has an annual average temperature of 53.1 Fahrenheit ranking 20th out of 54.
46. The year 1927 ranking 46th in number of whales caught, with a CO2 emission of 0.325 gigatons ranking 31st out of 69, has an annual average temperature of 52.5 Fahrenheit ranking 24th out of 54.

47. The year 1978 ranking 47th in number of whales caught, with a CO2 emission of 0.5833.32 gigatons ranking 18th out of 69, has an annual average temperature of 53.3 Fahrenheit ranking 18th out of 54.
48. The year 1940 ranking 48th in number of whales caught, with a CO2 emission of 0.40625 gigatons ranking 27th out of 69, has an annual average temperature of 53.8 Fahrenheit ranking 13th out of 54.
49. The year 1912 ranking 49th in number of whales caught,with a CO2 emission of 0.153 gigatons ranking 43rd out of 69, has an annual average temperature of 56.6 Fahrenheit ranking 1st out of 54.
50. The year 1913 ranking 50th in number of whales caught, with a CO2 emission of 0.28125 gigatons ranking 35th out of 69.
51. The year 1914 ranking 51st in number of whales caught, with a CO2 emission of -0.325 gigatons ranking 62nd out of 69, has an annual average temperature of 53.2 ranking 19th out of
52. The year 1911 ranking 52nd in number of whales caught, with a CO2 emission of 0.0625 gigatons ranking 49th out of 69, has an annual average temperature of 51.2 Fahrenheit ranking 37th out of 54.
53. The year 1979 ranking 53rd in number of whales caught, with a CO2 emission of 0.5625 gigatons ranking 19th out of 69, has an annual average temperature of 52.7 Fahrenheit ranking 23rd out of 54.
54. The year 1915 ranking 54th in number of whales caught, with a CO2 emission of -0.0625 gigatons ranking 53rd out of 69, has an annual average temperature of 53.2 Fahrenheit ranking 19th out of 54.
55. The year 1924 ranking 55th in number of whales caught, with a CO2 emission of 0.025 gigatons ranking 51st out of 69, has an annual average temperature of 51.5 Fahrenheit ranking 34th out of 54.

56. The year 1980 ranking 56th in number of whales caught, with a CO_2 emission of -0.125 gigatons ranking 58th out of 69, has an annual average temperature of 52.8 Fahrenheit ranking 22nd out of 54.
57. The year 1981 ranking 57th in number of whales caught, with a CO_2 emission of -0.4 gigatons ranking 63rd out of 69, has an annual average temperature of 54.3 Fahrenheit ranking 11th out of 54.
58. The year 1945 ranking 58th in number of whales caught, with a CO_2 emission of -0.890625 gigatons ranking 69th out of 69, has an annual average temperature of 50.5 Fahrenheit ranking 44th out of 54.
59. The year 1022 ranking 59th in number of whales caught, with a CO_2 emission of 0.153 gigatons ranking 43rd out of 69, has an annual average temperature of 51.2 Fahrenheit ranking 37th out of 54.
60. The year 1982 ranking 60th in number of whales caught, with a CO_2 emission of -0.166666 gigatons ranking 59th out of 69, has an annual average temperature of 51.0 Fahrenheit ranking 39th out of 54.
61. The year 1923 ranking 61st in number of whales caught, with a CO_2 emission of 0.4375 gigatons ranking 26th out of 69, has an annual average temperature of 50.8 Fahrenheit ranking 41st out of 54.
62. The year 1931 ranking 62nd in number of whales caught, with a CO_2 emission of -0.4375 gigatons ranking 64th out of 69, has an annual average temperature of 49.5 Fahrenheit ranking 50th out of 54.
63. The year 1910 ranking 63rd in number of whales caught, with a CO_2 emission of 0.15 gigatons ranking 44th out of 69, has an annual average temperature of 54.5 Fahrenheit ranking 9th out of 54.
64. The year 1983 ranking 64th in number of whales caught, with a CO_2 emission of -0.125 gigatons ranking 58th out of 69,

has an annual average temperature of 53.5 Fahrenheit ranking 16th out of 54.

65. The year 1941 ranking 65th in number of whales caught, with a CO2 emission of 0.09378 gigatons ranking 46th out of 69, has an annual average temperature of 51.5 Fahrenheit ranking 34th out of 54.
66. The year 1920 ranking 66th in number of whales caught, with a CO2 emission of 0.5 gigatons ranking 23rd out of 69, has an annual average temperature of 51.3 Fahrenheit ranking 36th out of 54.
67. The year 1984 ranking 67th in number of whales caught, with a CO2 emission of 0.65 gigatons ranking 15th out of 69, has an annual average temperature of 51.0 Fahrenheit ranking 39th out of 54.
68. The year 1919 ranking 68th in number of whales caught, with a CO2 emission of -0.5 gigatons ranking 67th out of 69, has an annual average temperature of 52.5 Fahrenheit ranking 24th out Of 54.
69. The year 1916 ranking 69th in number of whales caught, with a CO2 emission of 0.25 gigatons ranking 38th out of 69, has an annual average temperature of 51.1 Fahrenheit ranking 38th out of 54.
70. The year 1985 ranking 70th in number of whales caught, with a CO2 emission of 0.65625 gigatons ranking 14th out of 69, has an annual average temperature of 51.9 Fahrenheit ranking 39th out of 54.
71. Then year 1942 ranking 71st in number of whales caught, with a CO2 emission of -0.025 gigatons ranking 52nd out of 69, has an annual average temperature of 49.7 Fahrenheit ranking 48th out of 54.
72. The year 1909 ranking 72nd in number of whales caught, with a CO2 emission of 0.125 gigatons ranking 45th out of 69, has an annual average temperature of 51.8 Fahrenheit ranking 31st out of 54.

73. The year 1943 ranking 73rd in number of whales caught, with a CO_2 emission of 0.0625 gigatons ranking 49th out of 69, has an annual average temperature of 52.2 Fahrenheit ranking 27th out of 54.
74. The year 1921 ranking 74th in number of whales caught, with a CO_2 emission of -0.45 gigatons ranking 65th out of 69, has an annual average temperature of 54.4 Fahrenheit ranking 10th out of 54.
75. The year 1944 ranking 75th in number of whales caught, with a CO_2 emission of 0.0625 gigatons ranking 50th out of 69, has an annual average temperature of 49.4 Fahrenheit ranking 51st out of 54.
76. The year 1918 ranking 76th in number of whales caught, with a CO_2 emission of -0.0625 gigatons ranking 55th out of 69, has an annual average temperature of 52.3 Fahrenheit ranking 26th out of 54.
77. The year 1908, ranking 77th in number of whales caught, with a CO_2 emission of -0.125 gigatons ranking 58th out of 69, has an annual average temperature of 50.1 Fahrenheit ranking 47th out of 54.
78. The year 1986 ranking 78th in number of whales caught, with a CO_2 emission of 0.312 gigatons ranking 34th out of 69, has an annual average temperature of 52.7 Fahrenheit ranking 23rd out of 54.
79. The year 1917 ranking 79th in number of whales caught, with a CO_2 emission of 0.153 gigatons ranking 43rd out of 69, has an annual average temperature of 50.7 Fahrenheit ranking 42nd out of 54.
80. The year 1907 ranking 80th in number of whales caught, with a CO_2 emission of 0.3125 gigatons ranking 32nd out of 69, has an annual average temperature of 52.4 Fahrenheit ranking 25th out of 54.
81. The year 1906 ranking 81st in number of whales caught, with a CO_2 emission of 0.125 gigatons ranking 45th out of 69, has

an annual average temperature of 51.2 Fahrenheit ranking 37th out of 54.

82. The year 1987 ranking 82nd in number of whales caught, with a CO2 emission of 0.625 gigatons ranking 16th out of 69, has an annual average temperature of 53.4 Fahrenheit ranking 17th out of 54.
83. The year 1904 ranking 83rd in number of whales caught, with a CO2 emission of 0.025 gigatons ranking 51st out of 69, has an annual average temperature of 51.8 Fahrenheit ranking 31st out of 54.
84. The year 1996 ranking 84th in number of whales caught, with a CO2 emission of 0.688 gigatons ranking 13th out of 69, has an annual average temperature of 54.3 Fahrenheit ranking 11th out of 54.
85. The year 1997 ranking 85th in number of whales caught, with a CO2 emission of 0.15 gigatons ranking 44th out of 69, has an annual average temperature of 53.4 Fahrenheit ranking 17th out of 54.
86. The year 1998 ranking 86th in number of whales caught, with a CO2 emission of -0.083332 gigatons ranking 56th out of 69, has an annual average temperature of 52.9 Fahrenheit ranking 21st out of 54.
87. The year 1905 ranking 87th in number of whales caught, with a CO2 emission of 0.15 gigatons ranking 44th out of 69,has an annual average temperature of 51.7 Fahrenheit ranking 32nd out of 54.
88. The year 2001 ranking 88th in number of whales caught, with a CO2 emission of 0.25 gigatons 38th out of 69, has an annual average temperature of 53.8 Fahrenheit ranking 13th out of 54.
89. The year 1999 ranking 89th in number of whales caught, with a CO2 emission of 0.333 gigatons ranking 30th out of 69, has an annual average temperature of 53.3 Fahrenheit ranking 18th out of 54.

90. The year 2003 ranking 90th in number of whales caught, with a CO_2 emission of 1.35 gigatons ranking 2nd out of 69, has an annual average temperature of 55.0 Fahrenheit ranking 7th out of 54.
91. The year 2000 ranking 91st in number of whales caught, with a CO_2 emission of 0.688 gigatons ranking 13th out of 69, has an annual average temperature of 53.7 Fahrenheit ranking 14th out of 54.
92. The year 2002 ranking 92nd in number of whales caught, with a CO_2 emission of 0.6 gigatons ranking 17th out of 69, has an annual average temperature of 52.0 Fahrenheit ranking 29th out of 54.
93. The year 2013 ranking 93rd in number of whales caught, with a CO_2 emission of 0.3125 gigatons ranking 32nd out of 69, has an annual average temperature of 53.3 Fahrenheit ranking 18th out of 54.
94. The year 1992 ranking 94th in number of whales caught, with a CO_2 emission of -0.6875 gigatons ranking 68th out of 69, has an annual average temperature of 53.6 Fahrenheit ranking 15th out of 54.
95. The year 1994 ranking 95th in number of whales caught, with a CO_2 emission of 0.15 gigatons ranking 44th out of 69, has an annual average temperature of 54.7 Fahrenheit ranking 8th out of 54.
96. The year 2007 ranking 96th in number of whales caught, with a CO_2 emission of 0.938 gigatons ranking 7th out of 69, has an annual average temperature of 53.8 Fahrenheit ranking 13th out of 54.
97. The year 2004 ranking 97th in number of whales caught, with a CO_2 emission of 1.291666 gigatons ranking 3rd out of 69, has an annual average temperature of 52.1 Fahrenheit ranking 28th out of 54.
98. The year 2005 ranking 98th in number of whales caught, with a CO_2 emission of 1.0 gigaton ranking 5th out of 69, has an

annual average temperature of 53.4 Fahrenheit ranking 17th out of 54.

99. The year 2006 ranking 99th in number of whales caught, with a CO2 emission of 1.0 gigaton ranking 5th out of 69, has an annual average temperature of 53.8 Fahrenheit ranking 13th out of 54.
100. The year 2010 ranking 100th in number of whales caught, with a CO2 emission of 1.75 gigatons ranking 1st out of 69, has an annual average temperature of 52.7 Fahrenheit out of 54.
101. The year 1989 ranking 101st in number of whales caught, with a CO2 emission of 0.312 gigatons ranking 34th out of 69, has an annual average temperature of 52.2 Fahrenheit ranking 27th out of 54.
102. The year 1903 ranking 102nd in number of whales caught, with a CO2 emission of 0.1875 gigatons ranking 41st out of 69, has an annual average temperature of 50.6 Fahrenheit ranking 43rd out of 54.
103. The year 1993 ranking 103rd in number of whales caught, with a CO2 emission of 0.24 gigaton ranking 39th out of 69, has an annual average temperature of 50.3 Fahrenheit ranking 46th out of 54.
104. The year 1988 ranking 104th in number of whales caught, with a CO2 emission of 0.8333.32 gigaton ranking 10th out of 69, has an annual average temperature of 53.3 Fahrenheit ranking 18th out of 54.
105. The year 2009 ranking 105th in number of whales caught, with a CO2 emission of -0.458.332 gigaton ranking 66th out of 69, has an annual average temperature of 54.
106. The year 2011 ranking 106th in number of whales caught, with a CO2 emission of 1.15 gigaton ranking 4th out of 69, has an annual average temperature of 51.8 Fahrenheit ranking 31st out of 54.

107. The year 2012 ranking 107th in number of whales caught, with a CO2 emission of 0.5 gigaton ranking 23rd out of 69, has an annual average temperature of 56.6 Fahrenheit ranking 1st out of 54.
108. The year 1902 ranking 108th in number of whales caught,with a CO2 emission of 0.0625 gigatons ranking 49th out of 69, has an annual average temperature of 52.1 Fahrenheit ranking 28th out of 54.
109. The year 2008 ranking 109th in number of whales caught, with a CO2 emission of 0.5625 gigaton ranking 19th out of 69, has an annual average temperature of 51.9 Fahrenheit ranking 30th out of 54.
110. The year 1991 ranking 110th in whales caught, with a CO2 emission of 0.5 gigaton ranking 23rd out of 69, has an annual average temperature 51.8 Fahrenheit ranking 31st out of 54.
111. The year 2014 ranking 111th in number of whales caught, with a CO2 emission of 0.25 gigaton ranking 38th out of 69, has an annual average temperature of 55.6 Fahrenheit ranking 5th out of 54.
112. The year 1995 ranking 112th in number of whales caught, with a CO2 emission of 0.5 gigaton ranking 23rd out of 69, has an annual average temperature of 53.9 Fahrenheit ranking 12th out of 54.
113. The year 2015 ranking 113th in number of whales caught, with a CO2 emission of -0.3125 gigaton ranking 53rd out of 69, has an annual average temperature of 56.3 Fahrenheit ranking 2nd out of 54.
114. The year 1900 ranking 114th in number of whales caught, with a CO2 emission of 0.125 gigaton ranking 45th out of 69, has an annual average temperature of 53.9 Fahrenheit ranking 12th out of 54.
115. The year 1901 ranking 115th in number of whales caught, with a CO2 emission of 0.078125 gigaton ranking 48th out

of 69, has an annual average temperature of 53.5 Fahrenheit ranking 16th out of 54.

116. The year 2017 ranking 116th in number of whales caught, with a CO2 emission of 0.46875 gigaton ranking 24th out of 69, has an annual average temperature of 56.1 Fahrenheit ranking 4th out of 54.
117. The year 2016 ranking 117th in number of whales caught, with a CO2 emission of -0.0625 gigaton ranking 55th out of 69, has an annual average temperature of 56.2 Fahrenheit ranking 3rd out of 54.

Let us proceed reading the information in the tables above:

1. The year 2016, first in less whales caught, with -0.0625 gigaton CO2 emitted ranking 55th out of 69 (from high to low), has 56.2 Fahrenheit ranking third highest annual average temperature.
2. The year 2017, second in less whales caught, with 0.46875 gigaton CO2 emitted ranking 24th out of 69 (from high to low), has 56.1 Fahrenheit ranking fourth highest annual average temperature.
3. The year 1901, third in less whales caught, with 0.07812 gigatons CO2 emitted ranking 48th out of 69 (from high to low), has 53.5 Fahrenheit ranking sixteenth highest annual average temperature.
4. The year 1900, fourth in less whales caught, with 0.125 gigaton CO2 Emitted ranking 45th out of 69 (from high to low), has 53,9 Fahrenheit ranking twelveth highest annual average temperature.
5. The year 2015, fifth in less whales caught, with -0.312 gigaton CO2 emitted ranking fifty-third out of 69 (from high to low), has 56.3 Fahrenheit ranking second highest annual average temperature.

6. The year 1995, sixth in less whales caught, with 0.5 gigaton CO2 emitted ranking twenty-third out of 69 (from high to low), has 53.9 Fahrenheit ranking twelveth highest annual average temperature.
7. The year 2014, seventh in less whales caught, with 0.25 gigaton CO2 emitted out of 69 (from high to low), has 55.6 Fahrenheit ranking fifth highest annual average temperature.
8. The year 1991, eighth in less whales caught with 0.5 gigaton CO2 emitted ranking twenty-third out of 69 (from high to low), has 51.8 Fahrenheit ranking thirty-first highest annual average temperature.
9. The year 2008, ninth in less whales caught with 0.5625 gigaton CO2 emitted ranking nineteenth out of 69 (from high to low), has 51.9 Fahrenheit ranking thirtieth highest annual average temperature.
10. The year 1902, tenth in less whales caught with 0.0625 gigaton CO2 emitted ranking forty-nineth out of 69 (from high to low), has 52.1 Fahrenheit ranking twenty-eighth highest annual average temperature.
11. The year 2012, eleventh in less whales caught with 0.5 gigaton CO2 emitted ranking twenty-third out of 69 (from high to low), has 56.6 Fahrenheit ranking first highest annual average temperature.
12. The year 2011, twelveth in less whales caught with 1.15 gigaton CO2 emitted ranking fourth out of 69 (from high to low), has 51.8 Fahrenheit ranking thirty-first highest annual average temperature.
13. The year 2009, thirteenth in less whales caught with -0.458.332 gigaton CO2 emitted ranking sixty-sixth out of 69 (from high to low), has 52.1 Fahrenheit ranking twenty-eighth highest annual average temperature.
14. The year 1988, fourteenth in less whales caught with 0.8333.32 gigaton CO2 emitted ranking tenth out of 69 (from high to

low), has 53.3 Fahrenheit ranking eighteenth highest annual average temperature.

15. 1993, fifteenth in less whales caught with 0.24 gigaton CO2 emitted ranking thirty-nineth out of 69 (from high to low), has 50.3 Fahrenheit ranking thirty-nineth highest annual average temperature.
16. The year 1903, sixteenth in less whales caught with 0.1875 gigaton CO2 emitted ranking forty-first out of 69 (from high to low), has 50.6 Fahrenheit ranking forty-third highest annual average temperature.
17. The year 1989, seventeenth in less whales caught with 0.312 gigaton CO2 emitted ranking thirty-fourth out of 69 (from high to low), has 52.2 Fahrenheit ranking twenty-seventh highest annual average temperature.
18. The year 2010, eighteenth in less whales caught with 1.75 gigaton CO2 emitted ranking first out of 69 (from high to low), has 52.7 Fahrenheit ranking twenty-third highest annual average temperature.
19. The year 2006, nineteenth in less whales caught with 1.0 gigaton CO2 emitted ranking filth out of 69 (from high to low), has 53.8 Fahrenheit ranking thirteenth highest annual average temperature.
20. The year 2005, twentieth in less whales caught with 1.0 gigaton CO2 emitted ranking filth out of 69 (from high to low), has 53.4 Fahrenheit ranking seventeenth highest annual average temperature.
21. The year 2004, twenty-first in less whales caught with 1.291666 gigaton CO2 emitted ranking third out of 69 (from high to low), has 52.1 Fahrenheit ranking twenty-eighth annual average temperature.
22. The year 2007, twenty-second in less whales caught with 0.938 gigaton CO2 emitted ranking seventh out of 69 (from high to low), has 53.8 Fahrenheit ranking thirteenth highest annual average temperature.

23. The year 1994, twenty-third in less whales caught with 0.15 gigaton CO2 emitted ranking forty-fourth out of 69 (from high to low) has 54.7 Fahrenheit ranking eighth highest annual average temperature.
24. The year 1992, twenty-fourth in less whales caught with -0.6875 gigaton CO2 emitted ranking sixty-eighth out of 69 (from high to low), has 53.6 Fahrenheit ranking fifteenth highest annual average temperature.
25. The year 2013, twenty-filth in less whales caught with 0.3125 gigaton CO2 emitted ranking thirty-second out of 69 (from high to low), has 53.3 Fahrenheit ranking eighteenth highest annual average temperature.
26. The 2002, twenty-sixth in less whales caught with 0.6 gigaton CO2 emitted ranking seventeenth out of 69 (from high to low), has 52.0 Fahrenheit ranking twenty-nineth highest annual average temperature.
27. The year 2000, twenty-seventh in less whales caught with 0.688 gigaton CO2 emitted ranking thirteenth out of 69 (from high to low), has 53.7 Fahrenheit ranking fourteenth highest annual average temperature.
28. The year 2003, twenty-eighth in less whales caught with 1.35 gigaton CO2 emitted ranking second out of 69 (from high to low), has 55.0 Fahrenheit ranking seventh highest annual average temperature.
29. The year 1999, twenty-nineth in less whales caught with 0.333 gigaton CO2 emitted ranking thirtieth out of 69 (from high to low), has 53.3 Fahrenheit ranking eighteenth highest annual average temperature.
30. The year 2001, thirtieth in less whales caught with 0.25 gigaton CO2 emitted ranking thirty-eightieth out of 69 (from high to low), has 53.8 Fahrenheit ranking thirteenth highest annual average temperature.
31. The year 1905, thirty-first in less whales caught with 0.15 gigaton CO2 emitted ranking forty-fourth out of 69 (from

high to low), has 51.7 Fahrenheit ranking thirty-second highest annual average temperature.

32. The year 1998, thirty-second in less whales caught with -0.083332 gigaton CO2 emitted ranking fifty-sixth out of 69 (from high to low), has 52.9 Fahrenheit ranking twenty-first highest annual average temperature.
33. The year 1997, thirty-third in less whales caught with 0.15 gigaton CO2 emitted ranking forty-fourth out of 69 (from high to low), has 53.4 Fahrenheit ranking seventeenth highest annual average temperature.
34. The year 1996, thirty-fourth in less whales caught with 0.688 gigaton CO2 emitted ranking thirteenth out of 69 (from high to low), has 54.3 Fahrenheit ranking eleventh highest annual average temperature.
35. The year 1904, thirty-filth in less whales caught with 0.025 gigaton CO2 emitted ranking fifty-first out of 69 (from high to low),has 51.8 Fahrenheit ranking thirty-first highest annual average temperature.
36. The year 1987, thirty-sixth in less whales caught with 0.625 gigaton CO2 emitted ranking sixteenth out of 69 (from high to low), has 53.4 Fahrenheit ranking seventeenth highest annual average temperature.
37. The year 1906, thirty-seventh in less whales caught with 0.125 gigaton CO2 emitted ranking forty-five out of 69 (from high to low), has 51.2 Fahrenheit ranking thirty-seventh highest annual average temperature.
38. The year 1907, thirty-eighth in less whales caught with 0.3125 gigaton CO2 emitted ranking thirty-second out of 69 (from high to low), has 52.4 Fahrenheit ranking twenty-filth highest annual average temperature.
39. The year 1917, thirty-nineth in less whales caught with 0.153 gigaton CO2 emitted ranking thirty-second out of 69 (from high to low), has 50.7 Fahrenheit ranking forty-second highest annual average temperature.

40. The year 1986, fortieth in less whales caught with 0.312 gigaton CO2 emitted ranking thirty-fourth out of 69 (from high to low), has 52.7 Fahrenheit ranking twenty-third highest annual average temperature.
41. The year 1908, forty-first in less whales caught with -0.125 gigaton CO2 emitted ranking fifty-eighth out of 69 (from high to low), has 50.1 Fahrenheit ranking forty-seventh highest annual average temperature.
42. The year 1918, forty-second in less whales caught with -0.0625 gigaton CO2 emitted ranking fifty-filth out of 69 (from high to low), has 52.3 Fahrenheit ranking twenty-sixth annual average temperature.
43. The year 1944, forty-third in less whales caught with 0.0623 gigaton CO2 emitted ranking fiftieth out of 69 (from high to low), has 49.4 Fahrenheit ranking fifty-first highest annual average temperature.
44. The year 1921, forty-fourth in less whales caught with -0.45 gigaton CO2 emitted ranking sixty-filth out of 69 (from high to low), has 54.4 Fahrenheit ranking tenth highest annual average temperature.
45. The year 1943, forty-filth in less whales caught, with 0.0625 gigaton CO2 emitted ranking forty-nineth out of 69 (from high to low), has 52.2 Fahrenheit ranking twenty-seventh highest annual average temperature.
46. The year 1909, forty-sixth in less whales caught, with 0.125 gigaton CO2 emitted ranking forty-filth out of 69 (from high to low), has 51.8 Fahrenheit ranking thirty-first highest annual average temperature.
47. The year 1942, forty-seventh in less whales caught, with -0.025 gigaton CO2 emitted ranking fifty-second out of 69 (from high to low), has 49.7 Fahrenheit ranking forty-eighth highest annual average temperature.
48. The year 1985, forty-eighth in less whales caught, with 0.65625 gigaton CO2 emitted ranking fourteenth out of 69 (from high

to low), has 51.9 Fahrenheit ranking thirtieth highest annual average temperature.

49. The year 1916, forty-nineth in less whales caught, with 0.25 gigaton CO2 emitted ranking thirty-eighth out of 69 (from high to low), has 51.1 Fahrenheit ranking thirty-eighth highest annual average temperature.
50. The year 1919, fiftieth in less whales caught, with -0.5 gigaton CO2 emitted ranking sixty-seventh out of 69 (from high to low), has 52.5 Fahrenheit ranking twenty-fourth highest annual average temperature.
51. The year 1984, fifty-first in less whales caught, with 0.65 gigaton CO2 emitted ranking fifteenth out of 69 (from high to low), has 51.0 Fahrenheit ranking thirty-nineth highest annual average temperature.
52. The year 1920, fifty-second in less whales caught, with 0.5 gigaton CO2 emitted ranking twenty-third out of 69 (from high to low), has 51.3 Fahrenheit ranking thirty-sixth highest annual average temperature.
53. The year 1941, fifty-third in less whales caught, with 0.09378 gigaton Co2 emitted ranking forty-sixth out of 69 (from high to low), has 51.5 Fahrenheit ranking thirty-fourth out of 69 (from high to low), has 51.5 Fahrenheit ranking thirty-fourth highest annual average temperature.
54. The year 1983, fifty-fourth in less whales caught, with -0.125 gigaton CO2 emitted ranking fifty-eighth out of 69 (from high to low), has 53.5 Fahrenheit ranking sixteenth highest annual average temperature.
55. The year 1910, fifty-filth in less whales caught, with 0.15 gigaton CO2 emitted ranking forty-fourth out of 69 (from high to low), has 54.5 Fahrenheit ranking ninth highest annual average temperature.
56. The year 1931, fifty-sixth in less whales caught, with -0.4375 gigaton CO2 emitted ranking sixty-fourth out of 69 (from

high to low), has 49.5 Fahrenheit ranking fiftieth highest annual average temperature.

57. The year 1923, fifty-seventh in less whales caught, with 0.4375 gigaton CO2 emitted ranking twenty-sixth out of 69 (from high to low), has 50.8 Fahrenheit ranking forty-first highest annual average temperature.
58. The year 1982, fifty-eighth in less whales caught, with -0.166666 gigaton CO2 emitted ranking fifty-nineth out of 69 (from high to low), has 51.0 Fahrenheit ranking thirty-nineth highest annual average temperature.
59. The year 1922, fifty-nineth in less whales caught, with 0.153 gigaton CO2 emitted ranking forty-third out of 69 (from high to low), has 51.2 Fahrenheit ranking thirty-seventh highest annual average temperature.
60. The year 1945, sixtieth in less whales caught, with -0.890625 gigaton CO2 emitted ranking sixty-nineth out of 69 (from high to low), has 50.5 Fahrenheit ranking forty-fourth highest annual average temperature.
61. The year 1981, sixty-first unless whales caught, with -0.4 gigaton CO2 emitted ranking sixty-third out of 69 (from high to low), has 54.3 Fahrenheit ranking eleventh highest annual average temperature.
62. The year 1980, sixty-second in less whales caught, with -0.125 gigaton CO2 emitted ranking fifty-eighth out of 69 (from high to low), has 52.8 Fahrenheit ranking twenty-second highest annual average temperature.
63. The year 1924, sixty-third in less whales caught, with 0.025 gigaton CO2 emitted ranking fifty-first out of 69 (from high to low), has 51.5 Fahrenheit ranking thirty-fourth highest annual average temperature.
64. The year 1915, sixty-fourth in less whales caught, with -0.0625 gigaton CO2 emitted ranking fifty-third out of 69 (from high to low), has 53.2 Fahrenheit ranking nineteenth highest annual average temperature.

65. The year 1979, sixty-filth in less whales caught, with 0.5625 gigaton CO2 emitted ranking nineteenth out of 69 (from high to low), has 52.7 Fahrenheit ranking twenty-third highest annual average temperature.
66. The year 1911, sixty-sixth in less whales caught, with 0.0625 gigaton CO2 emitted ranking forty-nineth out of 69 (from high to low), has 51.2 Fahrenheit ranking thirty-seventh highest annual average temperature.
67. The year 1914, sixty-seventh in less whales caught, with -0.325 gigaton CO2 emitted ranking sixty-second out of 69 (from high to low), has 53.2 Fahrenheit ranking nineteenth highest annual average temperature.
68. The year 1913, sixty-eighth in less whales caught, with 0.28125 gigaton CO2 emitted ranking thirty-filth out of 69 (from high to low), has 51.3 Fahrenheit ranking thirty-sixth highest annual average temperature.
69. The year 1912, sixty-nineth in less whales caught, with 0.153 gigaton CO2 emitted ranking forty-third out of 69 (from high to low), has 56.6 Fahrenheit ranking first highest annual average temperature.
70. The year 1940, seventieth in less whales caught , with 0.40625 gigaton CO2 emitted ranking twenty-seventh out of 69 (from high to low), has 53.8 Fahrenheit ranking thirteenth highest annual average temperature.
71. The year 1978, seventy-first in less whales caught, with 0.5833.32 gigaton CO2 emitted ranking eighteenth out of 69 (from high to low), has 53.3 Fahrenheit ranking eighteenth highest annual average temperature.
72. The year 1927, seventy-second in less whales caught with 0.325 gigaton CO2 emitted ranking thirty-first out of 69 (from high to low), has 52.5 Fahrenheit ranking twenty-fourth highest annual average temperature.
73. The year 1925, seventy-third in less whales caught, with 0.0625 gigaton CO2 emitted ranking forty-nineth out of 69 (from

high to low), has 53.1 Fahrenheit ranking twentieth highest annual average temperature.

74. The year 1926, seventy-fourth in less whales caught, with -0.09375 gigaton CO2 emitted ranking fifty-seven out of 69 (from high to low), has 53.9 Fahrenheit ranking twelveth highest annual average temperature.
75. The year 1977, seventy-filth in less whales caught, with 0.525 gigaton CO2 emitted ranking twenty-second out of 69 (from high to low), has 53.2 Fahrenheit ranking nineteenth highest annual average temperature.
76. The year 1928, seventy-sixth in less whales caught, with -0.025 gigaton CO2 emitted ranking fifty-second out of 69 (from high to low), has 50.7 Fahrenheit ranking forty-second highest annual average temperature.
77. The year 1932, seventy-seventh in less whales caught, with -0.3125 gigaton CO2 emitted ranking fifty-third out of 69 (from highest to lowest), has 48.2 Fahrenheit ranking fifty-fourth highest annual average temperature.
78. The year 1946, seventy-eighth in less whales caught, with 0.375 gigaton CO2 emitted ranking twenty-eighth out of 69 (from high to low), has 52.0 Fahrenheit ranking twenty-nineth highest annual average temperature.
79. The year 1976, seventy-nineth in less whales caught, with 0.96875 gigaton CO2 emitted ranking sixth out of 69 (from high to low), has 51.6 Fahrenheit ranking thirty-third highest annual average temperature.
80. The year 1933, eightieth in less whales caught, with 0.1873 gigaton CO2 emitted ranking sixth out of 69 (from high to low), has 50.5 Fahrenheit ranking forty-fourth highest annual average temperature.
81. The year 1975, eighty-first in less whales caught, with 0.03125 gigaton CO2 emitted ranking thirty-second out of 69 (from high to low), has 50.8 Fahrenheit ranking forty-first highest annual average temperature.

82. The year 1935, eighty-second in less whales caught, with 0.1875 gigaton CO2 emitted ranking forty-first out of 69 (from high to low), has 51.9 Fahrenheit ranking thirtieth highest annual average temperature.
83. The year 1929, eighty-third in less whales caught, with 0.275 gigaton CO2 emitted ranking thirty-sixth out of 69 (from high to low), has 49.3 Fahrenheit ranking fifty-second highest annual average temperature.
84. The year 1934, eighty-fourth in less whales caught, with 0.3123 gigaton CO2 emitted ranking thirty-third out of 69 (from high to low), has 55.3 Fahrenheit ranking sixth highest annual average temperature.
85. The year 1972, eighty-filth in less whales caught, with 0.71875 gigaton CO2 emitted ranking twelveth out of 69 (from high to low), has 52.3 Fahrenheit ranking twenty-sixth high annual average temperature.
86. The year 1939, eighty-sixth in less whales caught, with 0.25 gigaton CO2 emitted ranking thirty-eighth out of 69 (from high to low), has 51.7 Fahrenheit ranking thirty -second highest annual average temperature.
87. The year 1947, eighty-seventh in less whales caught, with 0.25 gigaton CO2 emitted ranking thirty-eighth out of 69 (from high to low), has 51.6 Fahrenheit highest annual average temperature.
88. The year 1974, eighty-eighth in less whales caught, with -0.041666 gigaton CO2 emitted ranking fifty-fourth out of 69 (from high to low), has 52.9 Fahrenheit ranking twenty-first highest annual average temperature.
89. The year 1948, eighty-nineth in less whales caught, with 0.265625 gigaton CO2 emitted ranking thirty-seventh out of 69 (from high to low), has 50.6 Fahrenheit ranking forty-third highest annual average temperature.
90. The year 1973, ninetieth in less whales caught, with 0.873 gigaton CO2 emitted ranking nineth out of 69 (from high to

low), has 51.4 Fahrenheit ranking thirty-fifth highest annual average temperature.

91. The year 1949, ninety-first in less whales caught, with -0.187 gigaton CO2 emitted ranking sixtieth out of 69 (from high to low), has 50.1 Fahrenheit ranking forty-seventh highest annual average temperature.
92. The year 1930, ninety-second in less whales caught, with -0.325 gigaton CO2 emitted ranking sixty-second out of 69 (from high to low), has 49.4 Fahrenheit ranking fifty-first highest annual average temperature.
93. The year 1950, ninety-third in less whales caught, with 0.75 gigaton CO2 emitted ranking eleventh out or 69 (from high to low), has 51.6 Fahrenheit ranking thirty-third highest annual average temperature.
94. The year 1952, ninety-fourth in less whales caught, with 0.09375 gigaton CO2 emitted ranking forty-seventh out of 69 (from high to low), has 51.1 Fahrenheit ranking thirty-eighth highest annual average temperature.
95. The year 1938, ninety-fifth in less whales caught, with -0.25 gigaton CO2 emitted ranking sixty-first out of 69 (from high to low), has 52.0 Fahrenheit ranking twenty-nineth highest annual average temperature.
96. The year 1936, ninety-sixth in less whales caught, with 0.375 gigaton CO2 emitted ranking twenty-eighth out of 69 (from high to low), has 52.2 Fahrenheit ranking twenty-seventh highest annual average temperature.
97. The year 1953, ninety-seventh in less whales caught, with 0.1875 gigaton CO2 emitted ranking forty-first out of 69 (from high to low), has 53.2 Fahrenheit ranking nineteenth highest annual average temperature.
98. The year 1971, ninety-eighth in less whales caught, with 0.6 gigaton CO2 emitted ranking seventeenth out of 69 (from high to low), has 50.5 Fahrenheit ranking forty-fourth highest annual average temperature.

99. The year 1970, ninety-nineth in less whales caught, with 1.15 gigaton CO2 emitted ranking fourth out of 69 (from high to low), has 51.6 Fahrenheit ranking thirty-third highest annual average temperature.
100. The year 1968, hundredth in less whales caught, with 0.65 gigaton CO2 emitted ranking fifteenth out of 69 (from high to low), has 50.3 Fahrenheit ranking forty-sixth highest annual average temperature.
101. The year 1961, one hundred and first in less whales caught, with 0.025 gigaton CO2 emitted ranking fifty-first out of 69 (from high to low), has 52.3 Fahrenheit ranking twenty-sixth high annual average temperature.
102. The year 1969, one hundred and second in less whales caught, with 0.875 gigaton CO2 emitted ranking eighth out of 69 (from high to low), has 52.0 Fahrenheit ranking twenty-nineth highest annual average temperature.
103. The year 1955, one hundred an third in less whales caught, with 0.625 gigaton CO2 emitted ranking sixteenth out of 69 (from high to low), has 49.6 Fahrenheit ranking forty-nineth highest annual average temperature.
104. The year 1956, one hundred and fourth in less whales caught, with 0.5 gigaton CO2 emitted ranking twenty-third out of 69 (from high to low), has 51.5 Fahrenheit ranking thirty-fourth highest annual average temperature.
105. The year 1951, one hundred and fifth in less whales caught, with 0.375 gigaton CO2 emitted ranking twenty-eighth out of 69 (from high to low), has 50.4 Fahrenheit highest annual average temperature.
106. The year 1937, one hundred and sixth in less whales caught, with 0.3123 gigaton CO2 emitted ranking thirty-third out of 69 (from high to low), has 51.4 Fahrenheit ranking thirty-filth highest annual average temperature.
107. The year 1967, one hundred and seventh in less whales caught, with 0.375 gigaton CO2 emitted ranking twenty-eighth out of

69 (from high to low), has 51.9 Fahrenheit ranking thirtieth highest annual average temperature.

108. The year 1954, one hundred and eighth in less whales caught, with 0.125 gigaton CO2 emitted ranking forty-fifth out of 69 (from high to low), has 52.8 Fahrenheit ranking twenty second highest annual average temperature.
109. The year 1962, one hundred and ninth in less whales caught, with 0.3125 gigaton CO2 emitted ranking thirty-second out of 69 (from high to low), has 50.6 Fahrenheit ranking forty-third highest annual average temperature.
110. The year 1957, one hundred and tenth in less whales caught, with 0.25 gigaton CO2 emitted ranking thirty-eighth out of 69 (from high to low), has 51.7 Fahrenheit ranking thirty-second highest annual average temperature.
111. The year 1966, one hundred and eleventh in less whales caught, with 0.541666 gigaton CO2 emitted ranking twentieth out of 69 (from high to low), has 52.4 Fahrenheit ranking twenty-fifth highest annual average temperature.
112. The year 1958, one hundred and twelfth in less whales caught, with 0.21875 gigaton CO2 emitted ranking fortieth out of 69 (from high to low), has 53.8 Fahrenheit ranking fifteenth highest annual average temperature.
113. The year 1965, one hundred and thirteenth in less whales caught, with 0.5 gigaton CO2 emitted ranking twenty-third out of 69 (from high to low), has 50.5 Fahrenheit ranking forty-fourth highest annual average temperature.
114. The year 1954, one hundred fourteenth in less whales caught, with 0.4380 gigaton CO2 emitted ranking twenty-fifth out of 69 (from high to low), has 51.7 Fahrenheit ranking thirty-second highest annual average temperature.
115. The year 1963, one hundred and fifteenth in whales caught, with 0.525 CO2 emitted ranking twenty-second out of 69 (from high to low), has 50.9 Fahrenheit ranking fortieth highest annual average temperature.

116. The year 1960, one hundred and sixteenth in less whales caught, with 0.53125 gigaton CO2 emitted ranking twenty-first out of 69 (from high to low), has 51.9 Fahrenheit ranking thirtieth highest annual average temperature.
117. The year 1964, one hundred and seventeenth in less whales caught, with 0.5625 gigaton CO2 emitted ranking nineteenth out of 69 (from high to low), has 48.3 Fahrenheit ranking fifty-third highest annual average temperature.

Based on the information above, we have the following:

1. The seven first in numbers of less whales caught, with the CO2 emitted between the twenty-third and fifty-fifth out of 69 ranking from highest to lowest, have the annual average temperature between the second and sixteenth highest out of 54.
2. From the eighth to fourteenth first in numbers of less whales caught, with the CO2 emitted between the fourth and sixty-sixth out of 69, - except the eleventh in number of less whales caught -, have the annual average temperature between the eighteenth and thirty-first highest out of 54.
3. From the fifteenth to forty-sixth first in number of less whales caught, with the CO2 emitted between the first, second, third, fifth and sixty-eighth out of 69, ranking from highest to lowest, have the annual average temperature between the seventh first and the fifty-first highest out of 54.
4. From the forty-seventh first to fifty-sixth in number of less whales caught, with the CO2 emitted between the fourteenth and sixty-seventh out of 69, from highest to lowest, have the annual average temperature between the ninth to fiftieth highest out of 54.
5. From the fifty-seventh first to sixty-eighth in number of less whales caught, with the CO2 emitted between the nineteenth

and sixty-ninth out of 69, have the annual average temperature between the eleventh to forty-fourth highest out of 54.

6. From the sixty-ninth first to seventy-fifth in number of less whales caught, with the CO_2 emitted between the eighteenth and fifty-seventh out of 69, have the annual average temperature between the first to twenty-fourth highest out of 54.
7. From the seventy-sixth first to one hundred and fifth in number of less whales caught, with the CO_2 emitted between the fourth and sixty-second out of 69, have the annual average temperature between the sixth to fifty-fourth highest out of 54.
8. From the one hundred and sixth first to one hundred seventeenth in number of less whales caught, with the CO_2 emitted between the nineteenth and forty-fifth out of 69, have the annual average temperature between the fifteenth to fifty-third highest out of 54.

Based on the above information, in general it seems to be correlation between less numbers of whales caught and high annual average temperatures. But in the the chapters above, it was said that more whales would absorb the extra CO_2 in atmosphere, extra CO_2 in atmosphere which is the main cause of global warming, climate change, thus lower the temperature. However it is important to note here that the task of getting the data on whales is not an easy task, though efforts are being made to reduce uncertainty. To realize this book we have used the data from the official whales organizations, from scientists, from reliable organizations.

Scientific research is generally targeting to reduce uncertainty. Among the causes leading to uncertainty: some countries could or can reduce their data and this can lead to some significant divergence between the data reported and the actual data. For instance "documents prove that the Soviet Union killed 48,477 humpback whales from 1948 until 1973. The country officially reported to the IWC that only 2,710 humpbacks had been killed."[112]

Another case is that: "airgun sounds can lead to a significant loss in communication range of Blue and Fin whales at 2000 km."[113] and lead to data inacuracy.

But all counts put together, it is relevant to spot out that whales are CO2 natural sink. And we can remember here as it was said in the introduction of this book, that: "in 1896 SVANTE ARRHENIUS, using his knowledge of physics and chemistry, put forwards the mathematical relationship between CO2 and atmospheric temperature. His observations led him to theorize - correctly - that human Industrial would increase the temperature of Earth's atmosphere, leading to the hypothesis about the relationship between CO2 concentrations and Earth's atmosphere temperature."[114]

And this Arrehenius's theory still "valid and forms the foundation of modern climate science. In 1896 Arrhenius calculated that doubling the amount of CO2 in the atmosphere would raise the Earth's temperature, a concept that is now central to understanding the greenhouse effect and climate change. Arrhenius's work demonstrated that CO2 along with other greenhouse gases, traps heat in the atmosphere, leading to a warming effect. Today, climate models which are far more sophisticated than Arrhenius's, still use the same basic

[112] Less off, Jonah swallows the whale, 11 J. Envtl. L. & Litig. 413 (1996)
Cited by Angela Lang. Detailed Discussion: The Global Protection of whales. Michigan State University College of Law 2002 United States.

[113] Umweltbundesamt
https://www.umweltbundesamt.

[114] Disposable City. Miami's Future on the shore of climate catastrophe.
MARIO ALEJANDRO ARIZA. Bold type Books
116 EAst 16th street, 8th Floor New York, NY 10003

principle: that increasing greenhouse gas concentrations particularly CO2, will lead to a Warming planet."[115]

Let us come back to sea otter. And about it, we can read: "according to available data, the sea otter world population in the early 1900 s,was estimated to be between 1,000 and 2,000 individuals, drastically reduced from a historical population of 150,000 to 300,000 due to extensive hunting for their fur. Today the population is slowly recovering with the most recent estimates placing the total number around 51,935 sea otters."[116]

---

[115] Al Overview
Is Svante Arrhenius's theory about the relationship between CO2 and atmospheric
Temperature still valid today?

[116] Al Overview
Sea Otter World population numbers from 1900 to present
Google search.

Chapter 4

What is already done to fight Global Warming/Climate Change?

Facing the global environmental threat, the world in general, and in particular the scientists, governments and organizations have not remained indifferent. Proposal of solutions, international commitments and meticulous studies, researches emerged to address this world's threat. It is important here to call to our attention that among the solutions proposed to fight global warming/climate change - whaling's regulation appears as one of the solutions to be exploited in the fight against global warming.

But given that "early attempt to regulate whaling proved unsuccessful, it is in 1946 that the International Whaling Commission (I.W.C) was established as the global body responsible for management of whales, under the International Convention for the Regulation of Whaling (I.C.R.W) to oversee the proper management of whale stocks and orderly development of the whaling industry. In 1982, I.W.C members decided to establish a moratorium, or a temporary ban, on commercial hunting to allow whale stocks to replenish. The moratorium came into force in 1986 and remains in place today. The commission also designated a sanctuary area in the Indian ocean in 1994 to address the need for long term protection for all the whale species for which the I.W.C has acknowledged conservation responsibility and for ecosystem restoration. In 1972, the United States Congress passed the Marine Mammal Protection Act (M.M.P.A), that makes it illegal for any person under legal jurisdiction of the United States to kill, hunt, injure or harass all species of marine mammals regardless of their population status. It also makes it illegal to import marine mammal or products made from them."[117]

"With these protective efforts, some species have begun to return to acceptable numbers but others, especially the right and blue whales, are still rare and endangered. Only the small mink Whale exists in populations great enough for sustainable whaling to be considered."[118]

And since the world became aware of global warming/climate change, contributions, insights are emerging from everywhere in terms of solutions to fight global warming/climate change. It is the case of "a Chilean Charity called the Foundation M.E.R.I that figured out the foundations for whale - based carbon market, installing early - warming acoustic buoys that will monitor the locations of whales and generate alternative routes for ships. It is believed to be the world's first project whales."[119]

This is just an example of what NGOs, intergovernmental organizations and other organizations like the Chilean NGO BLUE Whale Center, Greenpeace, Sea Shepherd, W.W.F, U.N.C.L.O.S with its Exclusive Economic Zone (E.E.Z), the International Fund for Animal Welfare (I.F.A.W), the United Nations Framework Convention on Climate Change (U.N.F.C.C.C), that serves as the framework for international cooperation on climate change adopted in 1992, shortly

---

[117] Angela Lang. Detailed Discussion: The Global Protection of Whales. Michigan State University College of Law 2002 United States.

[118] Idem.

[119] Can whales really save the world from global warming? - study - The Jerusalem Post
Despite past studies proving whales can absorb Carbon Dioxide, this study looks at where they produce more than they take.
By DANIELLE GREYMAN - KENNARD Published: June 9, 2023 21:18

after the first assessment report by the Intergovernmental Panel on climate change (I.P.C.C) in 1990.

It is as part of U.N.F.C.C.C that in Kyoto Protocol according to Wikipedia #2 that "industrialized nations agree under the Convention to support climate change activities in developing countries by providing financial support for action on climate change. It is also as part of U.N.F.C.C that the United Nations Paris Climate adopted in 2015 to combat climate change. It aims to limit the global temperature increase to well below 2 degree Celsius above pre-industrial levels and ideally to 1.5 degree Celsius. The agreement includes commitments from all major emitting countries to reduce their climate pollution and strengthen those commitments over time. It also provides a framework for countries to transparency and report progress."[120]

The business's world on its side, found some alternatives to whale's blubbers, and "due to the creation of alternative resources, whale's blubber is no longer needed for oil."[121]

Though whale's hunting is not only about its blubber, the whale's demand won't be the same as before and "most of whale's hunting today is primarily after its meat for human consumption. Although unauthorized commercial whaling is likely to continue for a long time, it appears that this trend is declining as certain countries continue to lose profits due to continually declining interests."[122]

Coming back to governments, 195 countries met in Paris under the United Nations Framework Agreement on climate change (U.N.F.C.C.C) and agreed on the need to counter the causes of global warming/climate change's threat. And "since June 2021, governments around the globe have created a promising wave of progress on new large - scale marine protection commitments and designations covering more than 3.4 million square kilometers (1.3 million square miles)

[120] United Nations Paris Climate Agreement.
Wikipedia # 2

[121] History of whaling/Why it Began and where it is now/Whale Facts.
https://www.Whalefacts.org/history-of-whaling

[122] Idem

across some of the world's most biologically significant waters. Over the past decade, and informed by the best available science, many countries have created large scale, highly and fully protected M.P.As that prohibit harmful human activities, such as industrial fishing. These efforts have helped to safeguard millions of square kilometers of ocean all around the globe. Although these actions are having a positive impact, marine biodiversity continues to decline at an unprecedented rate."[123]

And though multiple efforts were made, in 2021, "the amount of planet-warming carbon dioxide in the atmosphere broke a record, 50 percent higher than the pre-industrial average before human began the widespread burning of oil, gas and coal in the late 19th Century. That is more carbon dioxide in the atmosphere than at any time in at least 4 million years, National Oceanic and Atmospheric Administration officials said."[124]

The amount of methane in the atmosphere too has been increasing as we can see in our previous chapter, chapter 2. But the fight against global warming/climate change kept it on going. For example in regard to methane, "until recently, little was known about where leaks were occurring, or the best way to fix them. In 2012 we kicked off a research series to better pinpoint leaks - and to find solutions. It is the largest body of peer - reviewed research on the issue. Today we have much better data on where the methane is coming from and how to prevent it. Ground-based measurement tools along with a growing number of satellites - including one launched in 2024 by our methane SAT subsidiary - will make it faster and cheaper than ever to locate, measure and reduce emissions. In fact the International Energy Agency estimates that worldwide the oil and gas industry can

[123] Connecting Marine Protected Areas Can Improve Ocean Health.
By linking their M.P.As, regions could build climate resilience, reverse biodiversity loss, and support communities
Issue Brief July 1, 2022Projects: Pew Bertarelli
Ocean Legacy & Mediterranean sea.

[124] Carbon Dioxide Levels Are Highest in the Human History. The New York Times. By: Henty Fountain. June 3, 2022

achieve a 75% reduction using technologies available today-two-thirds of it at no net cost. The good news: in 2023, the US Environmental Protection Agency unveiled historic rules to limit methane pollution from oil and gas sources. The safeguards are a vital win for the climate clean air and public health."[125]

Keeping on in how to cut too sharply the emissions, the Intergovernmental Panel on Climate Change (I.P.C.C), a United Nations body established to assess scientific Information about Climate Change and its potential impacts "first says that the world would have to triple or even quadruple the share of clean energy that it uses by 2050 - and keeps scaling it up thereafter. Second, we'd have to get dramatically more efficient at using energy in our home, buildings and cars."[126]

And when we look at emissions of CO2 per sectors, for example "the largest source of emissions of carbon dioxide the most common greenhouse gas, in the United States is the transportation sector. Emissions from transportation surpassed emissions from the electric power sector five years ago and now constitute two-fifths of domestic emissions from burning fossil fuels. In 2021, CO2 emissions in the transportation sector were 6 percent less than they were in 2005. The decline in emissions from transportation has contributed to a drop of about 20 percent in total CO2 emissions in the United States since 2005.[127]

The average production-weighted carbon dioxide emissions for new model year (MY) 2021 light -duty vehicles was 348 grams per

---

[125] Methane: A Crucial Opportunity in the Climate fight - Environment Defense Fund.
Copyright © 2024

[126] How to stop global warming in 7 steps - Vox
By Brad Plumer/@bradplumer/brad@vox.com/ Nov 2, 2014, 12:51 pm EST

[127] Environmental Protection Agency. Revised 2023 and Later Model Year Light-Duty Vehicle Greenhouse Gas Emissions Standards: Regulatory update, EPA-420-F-21-077 (December 2021), https://tinyurl.com/27 Caws 5n (https://tinyurl.com/27 caws 5n) (PDF). Quoted from Emissions of Carbon Dioxide in the Transportation Sector. Congressional Budget Office

mile, down from 681 grams per mile for new vehicles in MY 1975- a decrease of 49%.[128]

And they will keep on working on the increase. For example the model year 2026 emission standards is projected to equivalent of 55 miles per gallon. That is right there, a "turnaround: in United States actually CO2 emissions decreased from 2005 to 2014, due in part to new energy-efficient technology and the use of cleaner fuels."[129]

As we said it already, Individually or in groups, work have been done to find solutions to fight against this world threat, global warming/climate change. And here below, we give some of them. Let start with Paul Hawken and his team with solutions proposed in the book: DRAWDOWN. THE MOST COMPREHENSIVE PLAN EVER PROPOSED TO REVERSE GLOBAL WARMING, that presents the following solutions:

> "REPOPULATING the Mammoth steppe
> ______________________________; To keep the planet cool,you want grasses in subpolar regions, not trees and you get grasses when you reintroduce herbivores. Buried in the Circumpolar region of the Arctic are 1.4 trillion of carbon, two times more than in all the forests on the planet. Permafrost covers 24 percent of the Northern Hemisphere. At warming of 1.5 degrees Celsius (2.7 degrees Fahrenheit), permafrost is thawing releasing significant amount of carbon and methane into the atmosphere. If melting continue beyond 2 degrees Celsius (3.6 degrees Fahrenheit), the emissions released from the permafrost will become a positive - feedback loop that

[128] FOT W# 1223, January 31, 2022: Average Carbon Dioxide Emissions for 2021 Model Year Light-Duty Vehicles at an All-time Low/...

[129] Ibidem: Environmental Protection Agency. Revised 2023 and Later Model Year Light-Duty Vehicle Greenhouse Gas Emissions Standars: Regulatory update. EPA-420-F-21-077 (December 2021), Https://tinyurl.com/27 Caws 5n (https://tinyurl.com/27 Caws 5n) (PDF). Quoted from Emissions of Carbon Dioxide in the Transportation Sector. Congressional Budget Office

> accelerates global warming. When horses, reindeer, musk oxen and other denizens of the frozen north push away the layer of snow and expose the turf underneath, the soil is no longer insulated by its snow cover and is 3 to 4 degrees Fahrenheit colder, a margin of safety the world needs while it transitions away from fossil fuels. With this, permafrost can be prevented, and this would be the single largest solution or potential solution of the one hundred described in this book."[130]

Continuing with the same book, presenting solutions to fight the global warming/climate change

> "PASTURE CROPPING
> ___________________________:increases the use of the land by double - cropping (grains and animals) while reducing impact and increasing carbon sequestration.
>
> ENHANCED WHEATHERING OF MINERALS
> ___________________________:natural rock weathering removes approximately 1 billion ton of atmospheric carbon dioxide annually. Mining and milling of silicate rocks containing olivine and applying the rock powder to land and water so that the soil, oceans and biota can act as "reactors "for accelerated weathering. To fully halt carbon dioxide accumulation through enhanced weathering would take a

[130] DRAWDOWN> THE MOST COMPREHENSIVE PLAN EVER PROPOSED TO REVERSE GLOBAL WARMING
Edited by Paul Hawken
PENGUIN BOOKS
An imprint of Penguin Random House LLC
375 Hudson Street New York, New York 10014
Penguin.com
Copyright © 2017 by Project Drawdown
"Reciprocity" by Jamine Benjus © 2017 Jamine Benjus. Published by arrangement with the author.

staggering effort, involving billions tons of mineral spread over a significant fraction of the earth's surface.

MARINE PERMACULTURE
______________________:may be one of the most extraordinary ways to reverse global warming by reforesting new kelp forest in the sea, hundreds of acres of underwater plantations situated offshore, floating forests in the middle of the ocean. They grow many times faster than trees or bamboo..

INTENSIVE SILVOPASTURE
______________________:when adding more cattle, planting different types of trees, and rotating the held more quickly, it could have a beneficial effect on land and climate.

ARTIFICIAL LEAF
______________________:is about doing artificial photosynthesis."[131]

Still in the same book,

"AUTONOMOUS VEHICLES
______________________:rather than using private owned cars that make spend trillions dollars per year in ownership, car's insurance and on top of that emit carbon dioxide; shifting to contemporary electric powered vehicles - which are at least four times more efficient than gasoline powered private owned cars. The contemporary car is not a driving machine but a parking machine for which 700 millions parking spaces have been built.

SOLID-STATE WAVE ENERGY

[131] Idem (Drawdown. The Most Comprehensive Plan....

_____________________________:the kinetic energy of oceans, which surge with roughly 80.000 terawatt hours of power, is extraordinary. It is a staggering amount of energy. A single terawatt is the equivalent of 1 trillion watts and sufficient to provide electricity to 33 millions U.S homes.

LIVING BUILDING
________________:for example, living buildings should grow food, produce net positive waste (a waste stream that nourishes living systems or land, create net-positive water, and generate more energy with renewable than they use.

It has to avoid all "re-listed" materials, such as PVC and formal de hyde. Living buildings make their greatest impact by producing more energy than they consume and offsetting all embodied carbon as well."[132]

We proceed with the same book.

DIRECT AIR CAPTURE
______________________:long term hopes are to use DAC (Direct Air Capture) and carbon dioxide storage to help achieve and maintain drawdown.

HYDROGEN - BORON FUSION
______________________________:is based on the idea to use the energy from the sun to power the earth. In 2015, a company that had been considered a maverick because of its unorthodox approach announced it had achieved one - half of the Holy Grail, the more difficult half, nick named "long enough. "The company, Tri Alpha Energy (TAE), had been secretive for most of its eighteen year history.

[132] Ibidem

SMART HIGHWAYS

__________________:Watt Way photovoltaic pavement, a French technology, that will allow to produce clean electricity for uses from LED highlighting to E.V charging, while improving tire grip and surface durability. A noise barrier lined with P.V panels may be another win-win solution."[133]

We are still proceeding with the proposed solutions from the same book.

HYPERLOOP

____________:the vision of the hyperloop, - is something like - seven hundreds miles of low - pressure tubes running up and down California.

It is based on a paper written by Elon Musk in 2013. Hyperloops, as proposed, are extremely efficient. The promise of a Hyperloop is speed; its virtue is how little energy it uses to move people and cargo. Estimates per passenger mile are 90 to 95 percent less than planes, trains, or cars. It is about thrusting you up to 760 miles per hour through steel conduits from San Francisco to Los Angeless in thirty-five minutes, for a cost of a bus ticket.

MICROBIAL FARMING

____________________:

INDUSTRIAL HEMP

_________________:is about using industrial hemp instead of cotton crop. The difference in impact between the two is remarkable. Cotton is the dirtiest crop in the world with respect to chemical use and is largely dependent on fossil fuel inputs. Hemp can be used as substitute for cotton.

[133] Ibidem

PERENNIAL CROPS

__________________:is the most effective way to sequester carbon in any agricultural system

Because they leave the soil intact. Perennial upland rice would minimize deforestation by farmers who cultivate rice for a few years and - burn techniques because of lack of fertility."[134]

Still with the proposed solutions from the same book.

A COW WALKS ONTO A BEACH

______________________________:when cows are fed with some sea weeds, like red Asparagapsis taxiformis it reduces methane production. The seaweed help cows digest more efficiently and methane the key waste product generated when cows process the food, dropped by 12 percent diet.

OCEAN FARMING

________________:Oceans farming require no fresh water, no deforestation, and no fertiliser - all significant downsides to land - based farming -they promise to be more sustainable than the most environmentally farms. Professor Ronald Osinga, at Wageningen University in the Netherlands, has calculated that a global network of sea vegetable farms totaling seventy thousand square miles - roughly the size of Washington State - could provide enough protein for the entire World population. The goal, according to chef Dan Barber, is to create a world where farms "restore instead of deplete "and below "every community to feed itself.

[134] Ibidem

SMART GRIDS
_____________:can reduce overall consumption while facilitating the shift away from centralized fossil fuel plants and their greenhouse gas emissions."[135]

To finish with the solutions proposed in the same book,

BUILDING WITH WOOD
____________________: first a unit of dry wood is 50 percent carbon, and that carbon is locked in while the wood is in use. Second, the process of producing those materials generate fewer greenhouse gas emissions than producing wood's alternatives. Cement, used in concrete and other building materials, is responsible for producing 5 to 6 percent of global emissions, twice as much as the aviation industry. Manufacturing steel beams requires six or twelve times more fossil fuel than producing laminated timber.

RECIPROCITY
___________: when trees are grouped together, it is for a reason. They are benefiting one another in some way. The "wood wide web "as Simard calls it, is an underground internet through which water, carbon, nitrogen, phosphorus and defense compounds are exchanged." That is how trees are healthy to breathe in carbon dioxide through photosynthesis."[136]

Now let us look at JAMES NIFFENEGGER (a researcher at the National Renewable Energy Laboratory (N.R.E.L) who is an author of a new study funded by the US Department of Energy's Water Power Technologies Office, who said that "a

[135] Ibidem

[136] Ibidem

> wide heap of rust - colored, brandy seaweed floats atop the caribbean sea when you sink it in deep sea, you can store the carbon dioxide it absorb from the air and water ". The study examined various methods to capture carbon dioxide from the air or ocean and permanently sequester or store it - or better yet, do both simultaneously. These techniques - called marine carbon sequestration and marine carbon dioxide removal - are almost as diverse as marine wild life: some involve farming or sinking seaweed, other inject captured carbon into deep sea rocks, and some capitalize on clever chemistry to remove carbon directly from the ocean. But almost all are relatively new and untested technologies, and their costs, environmental impacts, and potential efficacity are still largely unstudied. Most marine carbon capture, sequestration, and removal technologies are still in the early stages of development."[137]

Still in this section of propositions to fight global warming/climate change, "a new Chinese solar project, costing $1.65 billions, has an annual out put of 5.7 billion kilowatt - hours (Kwh) - half that of an average nuclear power plant,"[138]

The list of what is already done to fight the Century global threat is long. For example, there are initiatives already started by governments, organizations, societies etc.. to fight against deforestation. And to the general trend of the emissions from the land use change that "was increasing notably between 2000 and 2020, growing by around a third, the new dataset almost perfectly reverse this trend, suggesting that L.U.C (land use change) emissions have actually declined by around a third since 2000. Over the past decade, L.U.C emissions

[137] Can we counter Climate change by dumping carbon in the ocean?:r/ Futurology, R/Futurology, 2 years ago Altbekannt.

[138] "News" Can Ocean Energy Power Carbon Removal?
Study Finds offshore Energy could Help the World Reach Critical Carbon Removal Goals (And a Bubbly, Soda - Like Method Might Be Our Best Bet)
February 1, 2024/By CAITLIN McDERMOTT - MURPHY/ contact media relations.

went from increasing by 1.8% per year to decreasing by 4% per year in the latest version of the G.C.P data."[139]

Coming to the improvement of wind energy and solar energy, "renewable - energy additions grew 17 percent with a record +- 600 G.W of solar, +- 125 G.W of wind, and near doubling of grid storage installations."[140]

Proceeding with initiatives already started to fight global warming/climate change we can talk about: 1. THE SEA ANCHOR CROP'S FARMING in which seaweed - "one of the fastest- growing plants in the World - uses photosynthesis to pull carbon from atmosphere and the water with some varieties capable of absorbing five times more carbon dioxide than land based plants. Seaweed and algae farming have been around for more than a Century. We have been doing that for a very long time "Niffenegger said. "Not in United States, but in Asia. While seaweed and algae farming, seaweed sinking and artificial up -welling are all relatively low - cost options for carbon dioxide capture and removal,they come with trade offs. And all require lots of energy and could cause severe environmental damage if they capture the amount of carbon needed to fight climate change."[141]

We proceed with the same author: "but other non - biological methods so-called electrochemical methods, with fewer environmental risks, are more expensive, but could be our best bet according to the study. They could remove up to 10 billion ton of carbon dioxide per year - potentially enough to bail out our sinking boat."[142]

---

[139] Major downward revision in land-use emissions over the past decade.
In: Global CO2 emissions have been flat for a decade, new data reveals - Carbon Brief.

[140] R.M.I
https://rmi.org
The Energy Transition in 2025: What to Watch For
January 8, 2025.

[141] "News" Can Ocean Energy Power Carbon Removal?
Study Finds offshore Energy Could Help the World Reach Critical Carbon Removal Goals (And Bubbly, soda-Like Method Might Be our Best Bet)
Feb 1,2024/ By CAITLIN McDERMOTT-MURPHY) Contact media relation.

[142] Idem ("News" Can Ocean Energy Power Carbon Removal?

With the global population rise impacting the need of more food and protein sources, this enterprise can be beneficial "not only in diversifying protein choices, but also reducing global emissions by up to 1.47 gigatonnes per year in 2050, comparable to removing 393 coal fired power plant annually. However, while these foods are common in some cultures, awareness and their use remain limited at the global scale and prices are sometimes prohibitively high for consumers. Some countries are starting to do this. Australia, for example, is investing AUD $70 million (US 45 million) in the Blue Economy Cooperative Research Center. It aims to develop industries to increase Australian seafood and marine renewable energy production."[143]

1. Though these initiatives are already in usage, their technologies are limited to relatively small scale "they require lots of energy and could cause severe environmental damage if they are to be applied to capture the amount of carbon needed to fight climate change."[144]
2. THE OFFSHORE WIND ENERGY is one of the "ready - to - implement solutions. And this potential though costly is more than theoretical, with investment in ocean already ramping up significantly. Global pledges in offshore wind have approximately doubled in the last four years bringing targeted deployment to up to 2000 G.W (enough to power approximately 1.5 billion homes annually) by 2050. And certain countries are leading by example: Norway, for instance, is home to the world's largest fully operational floating offshore wind park, Hywind Tampen. However, offshore

[143] 7 Ocean-based Solutions to Climate Change/World Resources Institute. The Ocean Can Play a Bigger Role in Fighting climate change than Previously Thought.

[144] Ibidem ("News" Can Ocean Energy Power Carbon Removal?

wind and marine energy are still not yet fully deployed at the scale needed to dramatically reduce global warming".)[145]

And further development and research are needed to fully realize the potential of these technologies."[146]

Let us recall here that while the primary contributor of the carbon dioxide emissions in US "is the transportation sector, the primary contributor to global carbon dioxide in the world today is the power industry, which accounts for a significant portion of emissions according to Statista. Following closely behind is the transportation sector."[147]

3. TO TARGET THIS SECTOR (transportation sector), new light - duty vehicles have made a long way with the Corporate Average Fuel Economy (CAFE) standard for new light-duty vehicles model year 2025 of 54.5 miles per gallon (mpg) both passenger cars and light-duty trucks."[148]

   But the Corporate Average Fuel Economy standard twenty years ago, "in 2005, was 20 miles per gallon."[149]

---

[145] Ibidem (7 Ocean-based Solutions to climate change/World Resources Institute. The Ocean Can Play a Bigger Role in Fighting climate change than Previously Thought).

[146] Al Overview.
To power enough carbon removal to limit global warming, offshore wind and marine energy.

[147] Al overview.
The main cause of CO2 emissions in U.S is transportation sector, what is the main contributor of CO2 emissions worldwide?

[148] Al Overview.
What is the CAFE standard for new light vehicles model year 2025? Google Search.

[149] Ibidem (Environmental Protection Agency.
Revised 2023 and Later Model Year Light-Duty Vehicles Greenhouse Gas Emissions standards: Regulatory update EPA-420-F-21-077 (December 2021) https://tinyurl.com/27 caws 5n (https://tinyurl.com/27 caws 5n (PDF)
Quoted from Emissions of Carbon Dioxide in the Transportation sector. Congressional Budget Office.

4. THE GRADING OIL AND GAS based on their methane intensity, aiming for low-methane gas too, "is already being implemented today, though it's still in relatively early stages and faces challenges."[150] The list of what is already being done might be long, but for us let stop here.

After chapter 4, where we presented different proposals of solutions to be done or what are already been done to fight global warming/climate change; here we are to chapter 5, where we present our own contribution to fight global warming/climate change.

---

[150] AI Overview.
Is the grading oil and gas based on their methane intensity to ends-up with low-methane gas already doable today?

Chapter 5

Tip To Boost Climate Resilience

Based on what is in chapter 4, it appears that a lot, about fighting against global warming, climate change, have been done or projected to be done. However the greenhouse gas's effects are still present and global warming, climate change still a world's threat.

For example, the global average temperature in January 2025 was 1.75 degrees Celsius[151], above the pre-industrial level. And even the CO2 emissions still alarming. Here is an example of that: while the year 2021's record of the CO2's amount in atmosphere, with 36 billions tons, was said the highest than at any time in at least four million years, according to National Oceanic and Atmospheric Administration officials, "in 2025, the global CO2 emissions are projected to be around 39 billion tons."[152]

And that is global warming, climate change expected to get worse. It is going from bad to worse. Yet there used to be a relative stability some years ago. According to Jeff Goodell's account of the Earth's history "the last time the Earth was hotter than it is today was

[151] According to the Copernicus Climate Change Service (C3S), cited in World Meteorological Organization (W.M.O)

[152] Al Overview.
The year 2021 broke the record in CO2 emission with 36 billion tons into the atmosphere, what is the CO2 emission in 2025?

at least 125,000 years ago, long before anything that resembled human civilization appeared."[153]

This leads us to talk a bit about the atmosphere homeostasis. As long as we are talking about global warming, climate change which are happening in the Earth's atmosphere, it is relevant to consider what it is happening in atmosphere. We will come back to the Earth atmosphere homeostasis, but let look first at what Liliane Voyé - a sociologist functionalist - says about the society: "society is a kind of total organism composed of various functions of living organisms: production, consumption, transport, communication, just like living organisms, social life evolves and it is this evolution whose laws must be established."[154]

Let come back to the Earth atmospheric homeostasis saying that if we understand that within many biological organisms, the case of human body, there is an autoregulation mechanism, homeostasis, which is a tendency toward a relative stable equilibrium between the organs.

In the same way, it is fair to consider that the Earth planet in general and in particular the Earth's atmosphere too, has "a natural autoregulation mechanism, largely driven by life on Earth, particularly the oceans and biosphere, which maintain stable conditions for the planet's climate and atmospheric chemistry over geological timescales, though human activity is overwhelming these systems now. These processes involve the carbon cycle (oceans absorbing CO2), biological processes (like plant photosynthesis), water cycles, and chemical feedback loops, working to keep Earth habitable, unlike other planets with unstable atmosphere, notes this article on SpringerLink.

[153] Op cit. THE HEAT WILL KILL YOU FIRST. Life and Death on A Scorched Planet.
By Jeff Goodell Little Brown and Company.
Hachette Book Group
1290 Avenue of the Americas, New York, NY 10104
Little Brown.comFirst Edition:July 2023.

[154] Voyé (1998), p. 171

These key atmospheric Autoregulation Mechanisms: Ocean Absorption, Photosynthesis, Water Cycle & Clouds, Chemical Buffering and Albedo (Reflectivity); have kept the Earth's atmosphere relatively stable for millions of years, creating optimal conditions for life, a concept known as planetary homeostasis. However, the rapid increase in greenhouse gases from human activities is now overwhelming these natural regulatory capacities, leading to climate change, as explained in this article on Biology Libre Texts."[155]

With this, especially if not aware that when the autoregulation mechanism is disrupted, no more responding the way it used to be, the planet is affected negatively and that is what is happening; the world might be worried, especially that we are not yet where we are suppose to be in regard to the 1.5 degree Celsius target, set in 2015, ten years ago, at the Paris climate change conference.

And today some might say that we are not even half way to reach the 1.5 degree Celsius target, or to reach the equilibrium. Thus fall in the group of the pessimists who prevail that there is nothing to do against this phenomena, it is a fate; quoting in example the January 2025's global average temperature, 1.75 degree Celsius above pre-industrial level, and the 2025's global emission into the atmosphere.

Contrary to that, it is quite understandable that the extra CO2 emitted in the atmosphere, that kept accumulated since the industrial revolution era, can last for many years. Seeing the piece of puzzle in this way can put us on the side of those who think that the fact that we don't reach the 1.5 degree Celsius target yet, does not mean that we will never reach it, thus drop the arms when it comes to fight against global warming, climate change.

But by contrary join the UN Secretary General Antonio Guterres, when he said "individual years pushing past the 1.5 degrees

[155] Atmosphere and Climate Regulation - Biology Libre Texts
Feb 20, 2022 - Life on Earth plays a critical role in regulating the atmosphere.

limit do not mean the long-term goal is shot. It means we need to fight even harder to get on track."[156]

What is important today is that, about the global threat, global warming, climate change; the world has not deserted, but has decided to stand firm against it. For that, the world focuses generally on two main areas: on one hand, to reduce as low as possible the greenhouse gas emissions, while encouraging the use of clean energy. It is in this way that there are efforts to decarbonize our energy systems to reduce the CO2 emissions.

We here warmly salute the energy efficiency and renewable energy's usage with the making of model light-duty vehicles with its reduction of the greenhouse emission per mile, with the projections showing hope to improve more.

Besides the light-duty vehicles's making, there are different more solutions already proposed either for clean energy's production, removing or absorbing the greenhouse gases from the atmosphere, or ocean. It is the case of a new chinese solar project, "costing $1.65 billion, has an annual out put of 5.7 billion kilowatt-hours (kwh)-half that of an average nuclear power plant."[157]

That is something to encourage, since it uses renewable energy sources provided by the Sun and that it can emit little to no greenhouse gases into the atmosphere. However let us look at the advantages and disadvantages of the solar energy.

"ADVANTAGES: renewable, low maintenance costs, diverse applications, technology development, cheaper than electricity bills.

"DISADVANTAGES: weather dependent, solar energy storage is expensive, uses a lot of space."[158]

---

[156] WORLD METEOROLOGICAL ORGANIZATION.
WMO confirms 2024 as warmest year on record at about 1.55 degree Celsius above pre-industrial level. Press Release 10 January 2025.

[157] Can we Counter Climate Change by Dumping Carbon in the Ocean?: r/ Futurology

[158] Advantages & Disadvantages of Solar Energy
Green Match. Mentioned in Forbes Independent. The Guardian

Coming back to the two main axis on which the world is streaming is efforts to fight global warming, climate change, it was said that in one hand, is the reduction as possible as low, the greenhouse gas emissions.

In the other hand, to remove or absorb the greenhouse gases from the atmosphere or oceans. And when it comes to the removal or absorption of greenhouse from the atmosphere or oceans, there are techniques - "called marine carbon sequestration and marine carbon dioxide removal - which are as diverse as marine wild life. The carbon dioxide removal's industry require energy to pull out the carbon dioxide from the atmosphere and ocean.

Another option that could, according to Niffenegger's new study, power enough carbon removal to limit global warming to that critical 1.5 degrees Celsius. There is he said, enough offshore wind and marine energy above and within U.S. waters to power the removal of 10 billion tons of carbon dioxide per year. That is huge. But not every technology designed to remove or capture and sequester marine carbon dioxide can get us to that number. And each the research team found came with tradeoffs. And almost all these techniques are relatively new and untested technologies and their costs, environmental impacts and potential efficiency are still largely unstudied. Most marine carbon capture sequestration, and removal technologies are still in the early stages of development."[159]

In the lines above, we said that the world is focused generally on two main axis:

To reduce as low as possible the greenhouse gas emissions, while encouraging the use of clean energy.

To remove or absorb the greenhouse gases from the atmosphere or oceans.

Last updates 4 April 2023
Written by Aris Vourvoulias.

[159] Ibidem (Can we counter climate change by dumping carbon in the ocean?: r/Futurology. R/Futurology, 2 years ago altbekannt

It is in this second axis (the removal or absorption of CO2) that we must consider our contribution in the fight against this world's threat, the global warming/climate change.

We are part of the group of those who attest pertinently, the case of the IMF, "since the role of whales is irreplaceable in mitigating and building resilience to climate change, their survival should be integrated into the objectives of 190 countries that in 2015 signed the Paris Agreement for combating climate risk."[160]

And many are of the same opinion. This is the case of Danille Greyman-Kenmard, Michelle Carrere, Francesco Bassetti, Elsa Cabrera, ect.

Although not everyone is unanimous on this issue, we can read however the following: "we should bring back the whales because there are many environmental and ecological benefits to it. But not for climate reasons,or at least, we shouldn't expect them to be a major climate solution. Yes, they might store a bit a carbon and that's a nice bonus. But their impact is likely to be small. Overstating whale's climate potential is not helpful and could be damaging. Creating a $1 trillion carbon removal market for whales with almost no evidence that whales could actually remove $1 trillion worth of carbon would be irresponsible. We should be putting that money into technologies and solutions that we know can reduce emissions."[161]

The same author says "I'm far from the first to make this point. Last year, a group of leading biodiversity scientists published a paper in Nature climate change. RESISTING CARBONIZATION OF ANIMALS AS CLIMATE SOLUTIONS. They look at the state of the research in much more detail, but reach similar conclusions: "ultimately, we strongly endorse trophic rewilding as an approach to

[160] Can Whales Really Save the World from global warming? - Study- The Jerusalem Post.
By Danielle Greyman-Kenmard. Published: June 9, 2023 21:18

[161] Maybe research that improves our understanding of whales role in carbon cycling would qualify here. No, we shouldn't be spending billions on it but it could be worth much smaller investments.

protecting and restoring biodiversity. While rewilding may support climate mitigation in some circumstances, we caution that unverified and inflated economic valuations, selective media reporting and a narrow management focus on carbon could result in perverse outcomes and reputational risk from failed carbon capture, and distract from the urgent need to reduce fossil fuel emissions. We can not afford to make this mistake:"[162]

However, contrary to that, "scientific research increasingly confirms the vital contributions of whales to the oceans, the planet and mankind. Whales are at the heart of the food chain that supports ocean life helping fish and many other species to thrive and reproduce. Whales also help to stabilize the Earth's atmosphere in the removal of carbon dioxide and the generation of oxygen."[163]

With little chance of contradiction, whales are vital if we want to boost the natural resilience in fighting global warming/climate change. Directly or indirectly, whales have an important role to play to fight the global warming/climate change.

Most of those who oppose the pro-whales- who support whales as an important key player in the fight against global warming/climate change- don't totally deny that whales do the intake of CO2. They rather claim that it is too small to impact the threat. Besides there are those who recognize that whales play a significant role in the ocean carbon cycle, but disagree when it comes to the percentage of CO2 presented as being absorbed by whales.

We can recall here that the pro whales, supporting whales as important to curb the global warming/climate change, claim that whales are responsible for storing 22% of Earth's total carbon.[164] Some

---

[162] Could whales be a solution to climate change? Probably only a very small one. The numbers don't live up to some of the hupe.
HANNAH RITCHIE Mar 24, 2025.

[163] How whales help ocean ecosystems, the climate and mankind.
stopkillingwhales.com

[164] ScienceDaily. Whales could be a valuable carbon sink, says scientists/ScienceDaily https://www.sciencedaily.com/releases/2022/12/221215120727.htm

claim more than 22% "they capture about 37 billion metric tons of carbon dioxide (that's an estimated 40 percent of all CO2 produced!) and produce at least 50 percent of all oxygen in our atmosphere."[165]

But the cons group, on the other hand, points out that "while whales do play a significant role in the ocean's carbon cycle, the claim that they absorb 22% of the world's CO2 is a simplification. The ocean as a whole absorbs approximately 22% of anthropogenic CO2 emissions.[166]

Proceeding with the cons on whales as important to curb global warming /climate change, on June 5, 2023, answering to the question: "Do whales increase the removal of carbon from the atmosphere? - the answer is -

Despite some hope that this would be the case, a new study has found the amount of potential carbon capture by whales is too little to....[167]

With what is in the lines above, do we still think that whales are important to boost natural resilience against global warming?climate change?

Of course, directly or indirectly whales have an important role to play to fight global warming? climate change. To elaborate on this we take four scenarios:

1. Let us start with even the worse scenario, in which whales have 0 percent CO2 intake, and the biological pump and the whale fall are included in this 22% CO2 absorbed by the ocean,

[165] Whales in the carbon cycle: can recovery remove...
People ask
How much CO2 do whales absorb?
Do whales absorb four times the amount of CO2 than the Amazon rainforest?

[166] Al Overview
Scientific evidence, proof that whales absorb 22% of the world CO2.

[167] Whales Not to Be Counted on as Climate Savers. In ScienceDaily.
Whales could be a valuable carbon sink Say Scientists/ScienceDaily.
Date: December 15, 2022.

though some claim that "the ocean absorbs about 30% of the carbon dioxide that is released in the atmosphere."[168]

And we already know, the "ocean is the most significant carbon dioxide absorber, acting as the planet's largest carbon sink. It absorbs approximately 25% of all CO2 emissions and absorbs 90% of the excess heat generated by those emissions. Forests are another major CO2 absorber, acting as the second-largest carbon sink after the oceans."[169]

But many researches estimate that forests absorb 25% of human carbon emissions. Other studies, like the one from NASA Jet Propulsion Laboratory (JPL) (.gov), suggest a higher percentage."[170]

In the same way, a study finds that "Forests worldwide absorb about 31% more carbon dioxide than previously thought, study finds."[171]

With what is in the lines above, though the exact percentage of CO2 that forests absorb seems to be debatable, most studies on forest's CO2 absorption converge between 25 to 30%. And we know that forest is second CO2 natural sink after ocean. If forest that is second natural CO2 - to believe many studies - absorbs 25 to 30%, it is not plausible that ocean being the first natural CO2 sink, is reduced to 22%.

Let came back to our first scenario in which we consider that the ocean absorbs 25 to 30% and the forest absorbs 25%. In this

---

[168] N.O.A.A Ocean Acidification.Feb 25, 2025.

[169] Climate Central
https://www.climatecentrale.org> climate.matters> ocean...
Ocean warming/Climate Central. What absorb CO2 the most?

[170] Al Overview
Scientific evidence or proof that forest worldwide absorb 25% of the CO2 worldwide - Google Search.

[171] Carec
Forests worldwide absorb about 31% more carbon dioxide than previously thought, study finds. Ecosystem And Ecology, Forest. October 30, 2024

scenario, with these two CO_2 natural sinks- if they are being taken care- the planet can be saved as they will be absorbing at least 50% of the emissions.

2. The second scenario in which we agree with those who oppose the pro-whales (who support whale as an important key player in the fight against climate change). Those opposing the pro-whales say that the whale's CO_2 intake is too small to impact the threat. But here we consider that-the too small to be 10%, meaning that through whale fall, amount of CO_2 are kept in the whale's body. Through feeding as well, it accumulates CO_2 and stores it until the dead body decomposes and sinks to the ocean floor or on the ocean surface. In this scenario we consider that amount of CO_2 as 10% of the world CO_2 absorbed.
   However, in a study answering to the question: how much CO_2 do whales absorb?
   "They capture about 37 billion metric tons of carbon dioxide (that's an estimated 40 percent of all CO_2 produced !)"[172]

Then in this scenario, we consider that-the ocean absorbs 25-30%
- Whales absorb 10%
- And forests absorb 25%

In this scenario, the ocean together with whales and forests absorb 60-65% of the total CO_2 worldwide.

3. The third scenario in which we consider the ocean, the phytoplankton and whales as part of a system that absorb 40% attributing this 40% mainly to phytoplankton. Because "extensive scientific evidence support the claim that

[172] Whales in the carbon cycle: Can recovery remove... People ask How much CO_2 do whales absorb? Do whales absorb four times the amount of CO_2 than the Amazon rainforest? January 4, 2025.

phytoplankton absorb approximately 40% of the world's CO_2"[173]

Then in this scenario, we consider the all set (ocean, whales and phytoplankton) as absorbing 40% of the CO_2 while the forests absorb 25 to 30%. This will make an absorption of 65 to 70% of CO_2.

4. The fourth scenario. In this scenario, we consider the ocean, phytoplankton included absorb 40%, while whales absorb 5% and forests absorb 30%. This is not my invention, as we can read: Estimates of CO_2 absorption by forests vary, with some studies suggesting that forests absorb 25% of human carbon emissions. Other studies, like the one from NASA Jet Propulsion Laboratory (J.P.L) (.gov), suggest a higher percentage.[174]

   In the same way "Forest globally absorb approximately 30% of the carbon emissions generated by human activities, like burning fossil fuels. This equates to roughly 2.6 billion tonnes of CO_2 absorbed annually. It is important to note that this absorption is partially offset by deforestation and other disturbances, but forests still act as a significant carbon sink."[175]

   In this last scenario, oceans and whales combined to forests absorb 75% of CO_2 worldwide.

   Scientific evidence or proof that forest worldwide absorbs 25% of the CO_2 worldwide -Google Search.

   In our third scenario, we talk about phytoplankton, which are microscopic organism in the ocean that absorbs 40% of CO_2 worldwide. Talking about phytoplankton, "while

[173] Al overview
Scientific evidence, proof that phytoplankton absorb 40% of the world CO_2 - Google Search.

[174] Forests Absorb Twice As Much As They Emit Each year.

[175] Al Overview
What is the percentage of CO_2 absorbed by the forest worldwide? Google Search.

phytoplankton biomass in the world's oceans amounts to only +- 1-2% of the total global plant carbon, yet these organisms fix between 30 and 50 billion metric..."[176]

"Phytoplankton use sunlight and nutrients such as phosphate, nitrate and calcium to create energy and grow, taking in carbon dioxide and releasing oxygen. Phytoplankton perform about 50 percent of all the photosynthesis on the planet. It is estimated that they have produced about half of all the earth's oxygen. Plankton are the foundation of the entire marine food chain. The health and productivity of all marine life ultimately depend on them."[177]

With what is in the paragraphs above we just want to highlight the vital role of phytoplankton when it comes to life in our planet. And numerous studies, including those from NASA (National Aeronautics and Space Administration) and other research institutions, have provided evidence for the role of phytoplankton in absorbing CO2"[178]

But in the pages above, though, in one hand researchers, the case of Hannah Ritchie,[179] rise concerns about the whale's contributions when it comes to CO2 absorption, alarming that whale's contribution to CO2 absorption is too small to

---

[176] National Institute of Health (NIH) (.gov)
https://pubmed.ncbi.nlm.nih.gov>...
By PG Falkowski. 1994. Cited by 704.

[177] State of the Planet. News from the Columbia Climate School
CLIMATE ECOLOGY
Plankton Are central to Life on Earth. How is Climate Change Affecting Them?
By Renee Cho
August 23,2023.

[178] Al Overview
Scientific evidence, proof that phytoplankton absorb 405 of the world CO2 - Google Search

[179] Could whales be a solution to climate change? Probably only a very small one.
HANNAH RITCHIE Mar 24, 2025.

> be taken in account; we should expect whales to be a major global warming/climate solution and advice to be reluctant to conclusions discouraging the whale's potential to fight global warming/climate change.

In the other hand, several scientific researches, studies claim with pertinence that whales play an irreplaceable role in mitigating carbon dioxide removal.

"Whales are the heart of the food chain that support ocean life helping fish and many other species to thrive and reproduce. Whales also help to stabilize the Earth's atmosphere in the removal of carbon dioxide and the generation of oxygen."[180]

In the same way we can read: "there is growing scientific evidence that whales play a significant role in capturing carbon from the atmosphere."[181]

Let us look at it: whale plays its role as a climate solution directly and indirectly. - Directly through feeding, whales store carbon in their bodies. When they die and sink, the carbon is kept in their bodies for years and released when the bodies decompose. This all process is called whales falls.

> -Indirectly, whales move up and down deep of the ocean to its surface. And they defecate a jumbo sized waste. Through this movement, the whale's waste and nutrients are being distributed through the ocean, making available by then, the food for different organisms including phytoplankton.
>
> It is relevant to recall that phytoplankton locate near the ocean's surface, to get the sun's rays. It is through this whale's movement that nutrient on the ocean surface can get to the deep parts of the ocean, and what is in the deep can reach to the surface of the ocean and circulate drifted by ocean

[180] Ho whales help ocean ecosystems, the climate and mankind. Stopkillingwhales.com

[181] Protecting Whales is climate positive. April 22,2021W.W.F Whales.

currents. This increase the ability of the ocean to absorb CO2, and combining with other elements, influence the ocean biodiversity positively.

Up to now, the question of how whales are irreplaceable in the carbon dioxide absorption, still not yet answered. To come to it, let see in the following lines, pages. The world has come to the evidence that to fix the climate change's crisis, we have to reduce and remove or absorb the excess of the CO2 emissions.

As I said it previously, we salute the reduction of CO2 through the making of vehicles with light CO2 energy emissions, or vehicles powered by energy emitting reduced amount of CO2, the making of electrical powered vehicles. The grading of methane gas too, brings a positive impact, quantifying and classifying methane emissions along the natural gas supply chain to indicate their environment performance.

Alongside the greenhouse gase's reduction, we have the removal, absorption of the CO2. And here there are two approaches :

1. THE CARBON REMOVAL OR ABSORPTION
   A natural, biological elimination process where by natural organic and inorganic materials and stored for long periods, often over centuries: green plants, soil, rock formations and ocean bodies can be used to facilitate carbon removal. Largely a natural process, carbon removal poses little or no-risk to the delicate balance of our ecosystems. But it takes a long time to sequester the gas. Therefore, on its own, carbon removal can not offset the carbon emission burden. The negative impacts of these manmade processes on climate change are too enormous to be adequately tackled by carbon removal alone.

2. CARBON CAPTURE
   "An artificial, mechanical driven by carbon capture technology and conducted in controlled environments using carbon sequestration equipment. Gas capture equipment collect CO2

which is then pumped into subterranean rock formations for storage.

As an emission control tool, the carbon capture using carbon sequestration equipment, is proving to be an effective tool to combat the negative impacts of global warming and climate change. However, like any artificial process, there are concerns with utilizing carbon capture technology for gas sequestration."[182]

Proceeding with the same author, "By comparison with carbon removal, carbon capture is a faster way to eliminate carbon dioxide from the environment and reduce its negative greenhouse gas effect. Through its various technologies, carbon capture can effectively dissipate carbon dioxide generated from industrial activities and help reverse the negative trend of climate change. While carbon capture appears to be the more effective strategy, a synergistic approach integrating natural carbon removal processes is the best way to achieve net zero emission targets."[183]

Contrary to the thesis above, "a recent study by the University of Oxford estimates that net-zero path-ways heavily dependent on carbon capture and storage (C.C.S) could cost at least $1 trillion more annually than scenarios involving renewable. Additionally, C.C.S is not as effective as phasing out fossil fuels in terms of limiting emissions. In a recent briefing to the CO2 Plenary Group, senior Economics in context Initiative researcher June Sekera explores the effectiveness, resource efficiency and co-impacts of two different approaches to carbon dioxide removal-mechanical and biological- and argues that bio-

[182] DX/IFS
(https://ifsolutions.com) Uncategorized (https://ifsolutions.com/category/uncategorized/)
Carbon capture vs carbon Removal: what is the difference?
February 21, 2023 Welcome.

[183] Idem

logical methods are superior to the mechanical methods on all three aspects. The briefing by Sekera was based on two years of research she conducted alongside a team of scientists and economists, the results of which were published in the journal P.L.O.S climate. Compared to the mechanical methods, the Biological Carbon Dioxide Removal (B.C.D.R) can be highly effective. The same I.P.C.C report estimates that with biological methods 0.86 Gt (gigaton) of carbon dioxide will be removed by 2030, and 4.19 Gt will be removed by 2100."[184]

Let us proceed with the same publication, "Biological sequestration methods with improved practices and ecosystem restoration have a much higher potential for carbon sequestration compared to the mechanical methods.

The evidence on the collateral impacts of the different methods of carbon removal clearly shows that the biological carbon dioxide removal (B.C.D.R) method has positive ecological impacts, enriching ecosystems and extending benefits to local communities, while the co-impacts of mechanical methods are overwhelmingly harmful. Sekera's analysis indicates that mechanical carbon dioxide removal (C.D.R) methods, can be 25 to 50 times more expensive than biological C.D.R methods. Sekera's research shows that the biological methods of carbon removal are effective, resource efficient and have low costs and positive co-impacts compared to the mechanical methods powered by fossil fuels that are ineffective, inefficient and have high costs and many negative co-impacts. In US over $11.7 billion went towards mechanical carbon capture projects from 2010-2020 and an additional $12 billion was included in the 2021 Infrastructure and jobs Act, but to date there has not been significant funding for biolog-

[184] Carbon Dioxide Removal: A comparative Analysis of Biological and Mechanical Approaches/Economics in context Initiative.
Global Development Policy Center Economics Context Initiative. Posted 1 year ago on Tuesday, December 12th, 2023 in Blog. Latest stories
Tagged: 2023, 2024, carbon dioxide,...
June-Sekera, Kaylee, Wei, Sekera, Sustainability, well being.

ical sequestration methods. Sekera strongly emphasizes the need for increased public investments in biological carbon sequestration."[185]

Based on June Sekera's analysis - cited in the lines above - and the research she conducted alongside a team of scientists the results of which were published in the journal P.L.O.S climate, it is quiet plausible to acknowledge that the biological, natural carbon dioxide elimination is more rational, effective and efficient.

Natural sinks are then important in the fight against global warming/climate change, if taken care effectively since "it's estimated that natural carbon sinks already absorb half of the human-produced carbon emissions which are driving the climate emergency."[186]

With respect to the preceding lines, it is rational that if the main CO2 natural sinks - including whales - are taken care of effectively, this will boost their resilience to fight against global warming/climate change; thus speed up to lower the negative effects of global warming/climate change.

And when it comes to CO2 natural sinks one of the most important carbon sink in our planet is the ocean, the seas, water body. When we name the world's biggest natural carbon sinks, naming in order from the biggest to the smallest, we have: "oceans, forests, soil and rock (carbon mineralization reaction), fungi, elephant."[187]

In fact the UN Climate -issues, ocean-impacts, says that "as the planet's greatest carbon sink, the ocean absorbs excess heat and energy released from rising greenhouse gas emissions trapped in the earth's system. Today the ocean has absorbed about 90 percent of the heat generated by rising emissions."[188]

---

[185] Idem: (Carbon Dioxide Removal: A comparative Analysis of Biological and Mechanical Approaches? Economics in context Initiative...

[186] Ibidem (World Economic Forum)
What are the World's biggest natural carbon sinks?
Jul 26, 2023.

[187] Idem (World Economic Forum)

[188] UN The United Nations
https://www.un.org > Climate-issues >

We still have the question, How whales play an irreplaceable key role in global warming/climate change's fight? But now we are heading toward the answer to that question.

We recall here that the oceans "depend on phytoplankton whether directly or indirectly. Phytoplankton produces most of the oxygen that we breath, more than trees and other land plants. Phytoplankton make up 98% of the ocean's food energy and are the foundation of virtually every marine food chain. Therefore, phytoplankton make most other ocean life possible. Without plankton, all ocean life would end leading to a terrestrial mass extinction rivaling the loss of the dinosaurus.[189]

But "like land plants, phytoplankton can not survive on carbon dioxide water and sunlight alone, they need nutrients for photosynthesis, especially iron and nitrogen. The most critical nutrient for photosynthesis is iron, yet it is the rarest substance in the ocean having a background concentration of just 3 parts per trillion.

Whale poo, of all things, is especially rich in iron and nitrogen. Since whale feces is fairly liquid and less dense than the sea water, it floats to the surface as plumes where it is readily available to the phytoplankton and in numerable organisms that rely on their nutrient plumes for survival. Whales are the primary species that fertilize phytoplankton around the oceans during their long migration routes, whilst sea currents and eddies distribute and mix the nutrients further.

According to a paper, published in the British Journal Proceeding of the Royal Society, Australian biologists did-some estimations on sperm whales in the Southern ocean-which represent just 3% of the global sperm whale population. And this can show that whales fertilize the world's oceans."[190]

Francesco Basetti too says: "due to both their size and swimming patterns, whales bothincrease the ocean's ability to absorb CO2 and

---

Ocean-impacts
How is the climate change impacting the World's ocean/United Nations.

[189] How Whales help ocean ecosystems, the climate and mankind. Stopkillingwhales.com

[190] Idem (How Whales help ecosystems, the climate and mankind.)

at the same time actually act as living carbon sinks that store CO_2 in their bodies."[191]

In the same way "scientific research increasingly confirms the vital contributions of whales to the oceans,the planet and mankind. Whales are at the heart of the food chain that supports ocean life helping fish and many other species to thrive and reproduce. Whales also help to stabilize the Earth's atmosphere in the removal of carbon dioxide and the generation of oxygen.

Whales are vital in maintaining the health and balance of ocean ecosystems and make a significant contribution to the Earth environment that directly affects each and everyone of us and future generations."[192]

No wonder why, rightly the "International Monetary Fund (IMF) study concludes that whale protection must now become a top priority in the global effort to tackle climate change." Since the role of whales is irreplaceable in mitigating and building resilience to climate change, their survival should be integrated into the objectives of the 190 Countries that in 2015 signed the Paris Agreement for combating climate risk, "the author write."[193]

Several researches, studies - on whales and global warming/climate change, quoted in chapter 3 of this book - advocate for whale's protection. And different protection's measures have been already taken and some being implemented. That is the case of the global moratorium established by the International Whaling Commission in 1986.

---

[191] Why whales are important For Carbon Sequestration. Francisco Basetti. August 27, 2020.

[192] Ibidem (How whales help ocean ecosystems, the climate and mankind. Stopkillingwhales.com

[193] Can whales really save the world from global warming?
-study- The Jerusalem Post
Despite past studies proving whales can absorb Carbon Dioxide, this study looks at where they produce more than they take. By DANIELLE GREYMAN-KENNARD
Published: June 9, 2023 21:18

In Protecting Whales to protect the planet, it says: "To preserve and protect whale populations, we must reduce the many dangers to whales in our seas. One way to do so would be to apply the UN - Program model for protecting forests. The UN - R.E.D.D Program gives incentive to countries to preserve their forests as a way of keeping carbon dioxide out of the atmosphere. "In similar way, we can create financial mechanisms to promote the restoration of the world's whale populations," said the report's authors...[194]

It is the case of the management of the Marine Protected Areas (M.P.A.S) too, among others. But although the efforts made to protect whales show some positive impacts, whales still threatened by direct or indirect threats.

In fact MIchael Gunther/W.W.F says: "despite a moratorium on commercial whaling and a ban on international trade of whale products, three countries - Iceland, Japan, and Norway - continue their commercial whale hunts. Over 1,000 whales a year are killed for such commercial purposes...."[195]

And we can recall here that even today, whale's recovery from its threat of extinction can still be a subject of debate. For example in curious.earth, New research indicates blue whale population recovery, as we can read: "Once thought to be almost extinct, the blue whale population is believed to be making a strong recovery and even suspected to be stabilizing. Whale population is now believed to be between 10,000-25,000 rising from approximately 2,000 at its lowest point. Despite the rise in population, the species is still classified as endangered by the International Union for Conservation of Nature (I.U.C.N)"[196]

---

[194] Protecting Whales to protect the planet.
Friday Jul 21, 2023 at 5:07

[195] World Wildlife Fund - W.WF - Worl-wildlife.org

[196] Curious.earth
New research indicates blue whale population recovery.
By Paddy Trower
May 8, 2024 // Biodiversity, climate, News, Oceans.

In How Many Whales Are Left In the World?, Kathryn Dueck says: "estimate suggest that there are at least 1.5 million whales left in the world. This estimate is incomplete, however as it does not cover all species. It is therefore impossible to know the exact number of whales left. Certain species are rarer than others. The blue whales has garnered much attention for both its massive and its endangered status. Approximately 10,000 to 25,000 of these gentle giants remain in the wild today, a huge decrease from the 250,000 individuals roaming the seas years ago. The North Atlantic Right Whale is in even worst shape, listed by the International Union for the Conservation of Nature as critically endangered. There are approximately 370 of them in the wild today. But even worst of all is the Baiji, a species of freshwater river dolphin, as scientists now believe that it is functionally extinct due to the back of recent verified sightins."[197]

We can also read that: "the critically endangered North Atlantic Right whale population has continued its slow increase from a low of 358 individuals in 2020, according to the most recent estimate from the North Atlantic Right Whale consortium. But scientists are hesitant to celebrate. The population of North Atlantic right whales - one of the world's most endangered large whale species - remains significantly smaller than it was a decade ago."[198]

We can go on and read: "the mere fact that several populations are increasing does not indicate that they are truly on the road to recovery. First, uncertainty surrounding historic and current population estimates for all species makes it difficult to know with confidence the extent to which populations of large whales have "recovered." In fact, no management agency has come up with a rigorous and objective definition for recovery in unambiguous manner. Now here in the E.S.A are there criteria presented to define recovery. Although the conservation biology community and the world Conservation Union

[197] How Many whales Are Left In the World.

[198] As the North Atlantic Right Whale Population Slowly Increases, Human Activity Remain a Serious Threat.
By New England Aquarium on Tuesday, October 22, 2024.

I.U.C.N - have recently made considerable progress in developing methods for classifying the degree to which a population is vulnerable to extinction, many of these criteria are not easily adapted to marine species. For the most part, our limited knowledge of large whale biology makes it extremely difficult to quantify the degree to which a population may go extinct in a specific period of time. The same author says: "on the other hand, there is adequate information to at least begin assessing some populations of large whales, including Bowhead, Gray, Humpback and Right Whales. Some of these populations are showing signs of recovery."[199]

All things considered, whales still need a special attention. Since whales are irreplaceable in their role to fight global warming/climate change, with regard to the harmful effects caused by global warming/ climate change, we should have zero commercial whale's hunting, and fight all form of whale's threats. We know that the threats are direct - commercial whaling - and indirect such as entanglement, ocean pollution from toxic chemicals and plastics, collision between ships and whales, etc.

We then need zero commercial whaling - as said in the lines above, as well as finding ways to eradicate or minimize the indirect threats.

And today, worrying about how to end the threats against whales, which will help to boost the resilience of the elements fighting against global warming/climate change, is more than a necessity. In his /her research's finding, Paddy says: "the findings of the research on blue whale populations offer hope that conservation efforts are having positive outcomes. The scale of the International effort has demonstrated that species recovery can be possible, and shows how scientists can work together to learn more about endangered species populations.

[199] Measuring Success in Conservation.
By LEAH GERBER (/AUTHOR/LEAH_GERBER)
DOUGLAS DEMASTER (/AUTHOR/DOUGLAS DEMASTER), SIMONA ROBERTS (/AUTHOR/SIMONA_ROBERTS), Assessing efforts to restore populations of marine mammals is partly a matter of epistemology: How do you know when enough is enough?
July-August 2000

This is backed up by other research, which has shown that conservation is effective at reducing global biodiversity losses. Both research studies show that species population decline can be slowed or even reversed with the right interventions."[200]

That is where I come in with my contribution, in this book. We know that countries, commercial whaling companies, etc, always find ways to go around whale's conservation measures, international laws, international whaling ban, and continue killing whales for commercial reasons.

But beyond all moral considerations, and knowing that commercial whaling is one of the major factor accelerating global warming/climate change, affecting the entire planet, regardless that this is caused by some particular countries, whaling companies, etc.

Faced with this situation, we propose to definitely put an end to whaling, by finding ways to insert a radioactive isotope in whales, making their meat becoming a poison when eating it, or find a way to make their meat uneatable, and useless for the uses whales are hunted for. This might be seen as a very strong measure, inhuman, even criminal, but it worth it as long as the world can have whales protected against hunting.

However studies, researches need to be done to make sure the radioactive isotope injected in whales, is very slightly enriched, or has reduced radioactive effects that can not harm the whales themselves, with no risks or little negligible risks associated with potential leakage and long-term negative environmental impacts. We can refer to the works of professor James Larkin and his team at the university of the WITWATERSRAND in South Africa, who have discovered a way to save rhinos against poaching. It is called the Risotope Project, and it employs nuclear technology that can be detected by radiation monitors at international borders. It is about combating rhino poaching by

[200] Curious.earth
New research indicates blue whale population recovery
By Paddy Trower
May 8, 2024 // Biodiversity, climate change, News, oceans.

implanting harmless radioactive isotopes in rhino horns. It works on two fronts says the professor James Larkin: first, we use people's natural aversion to radioactive materials. So, we put a small quantity into the horn and tell the world, "Hey, look, rhino horn from Southern Africa is going to be radioactive now '. We rely on people's fear of radiation to scare them away and make them realize that the horn is no longer valuable. Secondly, by making the horn slightly radioactive, it becomes easier to detect as it is transported around the world. The quantity we've chosen is safe for the animals but sufficient to trigger detection alarms.... There is now a much larger team trained to detect rhino horn, increasing from a few hundred to several thousand people worldwide capable of intercepting the horns."[201]

With what is said in the lines above, why can't the world set up a research team, similar to James Larkin's team from the university of Witwatersrand to work on the possibility to save whales from hunting's threats? This team will be associated with a team composed with experts in whale's conservation, veterinarians and experts on environment in the world and find ways to use the finding of professor James Larkin and his team, to address the whale's endangered's threat effectively, rationally and efficiently.

This project will have the following targets:

To insert a harmless detectable radioactive isotope in whale, making it possible for the ship's pilots in cruse, to receive signals whenever a whale is around perimeters which can cause collision between the ship and whale. This will allow the ship's pilot to change the trajectory to avoid collision with whales.

To insert the radioactive isotope, making it possible to a trained team to see, observe and monitor the movements of whales in order to know whenever a whale is endangered by fishing nets or lines or something similar. This will allow to a trained team to go in - using appropriate engines - and assist whale by cutting off the nets, lines

[201] Biz News
Get context Know more.
Prof James Larkin

from whale to free it from that entanglement, instead of leaving whale struggle with it for several days or months until death.

Come to the point that the material implanted in whale can react and send signals in the monitoring computers set in an advanced monitoring office, whenever a gun capable of killing whales is brought in, at a certain perimeter around whale's zone. This will make it possible to monitor the whale's killing closely and possibly prosecute the culprit, potential whale's killer.

Alongside, we can count on the assumption that people will come to believe that whale's meat or other parts of whale are either no more eatable for the case of meat, or useless for the other parts of whales that people hunt whales for. With these we will be addressing both the direct whale's threat (the killing of whales), and the major indirect threats (the entanglement with fishing nets, the collisions between ships and whales, thus starting a new era of whale's re-bouncing.

Since for the case of rhinoceros, the isotope is inserted in the horn, which horn is made up of keratin, the interdisciplinary expert's team will find out whether the isotope can be put in the part of whale's body made up of keratin, for the case of baleen whale, or in the appropriate spot for the case of other types of whale.

Without minimizing the other threats of whale, like sea pollution, plastics in the ocean, which however need to be addressed as well; by using my proposal, the whale's spectacular re-bouncing will become a reality, thus the climate change resilience will be certain. With this, we reach at the conclusion of this book.

Conclusion

In chapter 1, you could read that our planet Earth as a system is one of the nine planets in solar system. This planet is made up of five subsystems, including a layer of gas, the atmosphere that is made up of Nitrogen (N2) 78% of air, Oxygen (O2) 21%, the inert gas Argon (Ar) 0.93% and trace amounts of Carbon dioxide (CO2), Neon (Ne), Helium (He), Nitrous oxide (No), Xenon (Xe), Ozone (Co), and Amonia (NH3)."[202]

When you read our introduction, you will find that we have used the structuro-functional method, system-based approach, composed of interacting parts in their function of maintaining the system.

And coming back to chapter 1, you can realize that in the subsystem atmosphere, it being observed a dramatic rise of carbon dioxide, of greenhouse gases, caused mainly by the industrial revolution that has led to an unexpected rise of temperatures, to global warming/climate change. To acknowledge that the rise of temperatures has something to do with the rise of CO2 amount in atmosphere, we can look at the case of Venus, "the hottest planet in solar system, with a temperature of 867 Fahrenheit due to its thick atmosphere with higher amount of carbon dioxide which traps heat inside it, and clouds made up of Sulfuric acid."[203]

Rhinos go radioactive in the fight against poaching - Prof James Larkin 7th March 2025 by Linda Van Tilburg.

[202] U.C.A.R/Center For Science Education S.C.I.E.D

[203] Solar system 101/National Geography.

What is in the Air/Center for Science Education.

Rightly so, about the rise of the temperature in atmosphere, about global warming/climate change, the scientific researches show that the dramatic rising amount of greenhouse gases, including CO2 is the main cause of the nowadays climate's crisis. And to address it we need, in one hand to cut and reduce the greenhouse gas's emission and in the other hand remove, absorb the excess of the greenhouse gases already in the atmosphere.

With regard to that, the world, in one hand is reducing the greenhouse gases, the carbon dioxide's emission by the making of vehicles emitting reduced, less greenhouse gases. In the other hand, several solutions: the removal, absorption (a natural or biologic process) or the capture (an artificial or mechanical process) of the excess of carbon dioxide existing.

And related to what is above, "Sekera research shows that the biological methods carbon removal are effective, resource efficient and have low costs and positive co-impacts compared to the mechanical methods powered by fossil fuels that are ineffective, inefficient and have high costs and many negative co-impacts."[204]

Going through the book you have realized that our contribution to address global warming/climate change, is in the natural, biological approach of carbon dioxide's absorption, removal done through carbon dioxide's natural sinks, including whales. Here we target on whale's rigorous conservation - alongside with the sea otters and forest efforts of conservation - to boost the climate change's resilience, by referring to the findings of professor James Larkin and his team, and use a radioactive isotope to be inserted in whales to protect them against killing, hunting, entanglement, collision between whales and ships etc.

It is evident that to address properly the climate change's crisis, we need a holistic approach. However, far from minimizing the other

[204] What are the advantages of the natural CO2 removal versus the other methods. Google-search

climate change's solutions, protecting whales is to protect the planet, is not a mere slogan, but a tip to use, because it is vital for the planet.

Therefore, in the same way that the world has a multi-billion dollar nuclear security infrastructure that exists,[205] as described by the department of Energy, the threat to life's extinction on Earth due to dramatic negative global warming/climate change's threats to come if we don't act, should prompt the world to establish a multi-billion dollar for climate change's resilience.

To end, let say that the world is rightly concerned about wars, lack of food, water shortage, poverty etc.., and is fighting to find resources to counter it. It is important to realize - and the world has become aware of this - that global warming/climate change, is leading us to the worst possible future.

On the following page we have some abbreviations and we provided their meaning to help the reader understand.

[205] There is a multi-billion-dollar nuclear security infrastructure, is there one for global warming/climate change?
Google Search.

Abbreviations

1. W.W.F: World wildlife Fund
2. E.S.A: Emotional Support Animal
3. I.U.C.N: International Union for Conservation of Nature
4. UN-R.E.D.D: United Nations Collaborative Program on Reducing Emissions from Deforestation and Forest Degradation
5. R.M.I Climate Intelligence Program: Rocky Mountain Institute.
6. D.A.C: Direct Air Capture
7. B.E.C.C.S: Bioenergy with carbon capture and storage.
8. I.P.C.C: Intergovernmental Panel on Climate Change.
9. T.A.E: Tri Alpha Energy.
10. E.E.S.I: Environmental and Energy Study Institute.
11. N.Y.D.F: New York Declaration on Forests.
12. N.C.E.I: National Centers for Environmental Information.
13. N.H.H.S: National Hydro-meteorological Service.
14. W.M.O: World Meteorological Organisation.
15. I.M.F: International Monetary Fund.
16. I.W.C: International Whaling Commission.
17. I.C.R.W: International Convention for the Regulation of Whaling.
18. M.M.P.A: Marine mammal Protection Act.
19. M.E.R.I: Monitoring, Evaluation, Reporting and Improvement.
20. U.N.C.L.O.S: United Nations Convention on the Law of the Sea.
21. I.F.A.W: International Fund for Animal Welfare.
22. U.N.F.C.C.C: United nations Framework Agreement on Climate Change.

23. E.E.Z: Exclusive Economic Zone.
24. N.O.A.A: National Oceanic and Atmospheric Administration.
25. L.U.C: Land Use Change.
26. AUD $: Australian dollar.
27. C.C.S: Carbon Capture Storage.
28. P.L.O.S: Public Library of Science.
29. B.C.D.R: Biological Carbon Dioxide Removal.

www.ingramcontent.com/pod-product-compliance
Lightning Source LLC
LaVergne TN
LVHW090321210926
848480LV00001B/316